# THE SS TERRA NOVA

WHALER, SEALER AND
POLAR EXPLORATION SHIP

# THE SS TERRA NOVA
## (1884–1943)

MICHAEL C. TARVER

Also by the same author
*The Man Who Found Captain Scott, Antarctic Explorer And War Hero: Surgeon Captain Edward Leicester Atkinson (1881–1929), DSO AM MRCS LRCP Royal Navy*

First published in 2006 by Pendragon Maritime Publications, Hillhead, Brixham, Devon, TQ5 0EZ

This edition published 2020

The History Press
97 St George's Place, Cheltenham,
Gloucestershire, GL50 3QB
www.thehistorypress.co.uk

© Michael C. Tarver, 2006, 2020

The right of Michael C. Tarver to be identified as the Author of this work has been asserted in accordance with the Copyright, Designs and Patents Act 1988.

All rights reserved. No part of this book may be reprinted or reproduced or utilised in any form or by any electronic, mechanical or other means, now known or hereafter invented, including photocopying and recording, or in any information storage or retrieval system, without the permission in writing from the Publishers.

British Library Cataloguing in Publication Data.
A catalogue record for this book is available from the British Library.

ISBN 978 0 7509 9408 8

Typesetting and origination by The History Press
Printed and bound by TJ Books, Padstow, Cornwall

 Trees for Life

# Contents

| | |
|---|---|
| Acknowledgements | 9 |
| Author's Note | 13 |
| Prologue | 15 |
| Foreword by HRH Princess Anne | 22 |
| Introduction by Sir Ranulph Fiennes | 23 |

1 Into the Evening of a Passing Age — 25
   Introduction
   The British Whaling Industry and its Ships
   The Shipyard of Alexander Stephen & Sons, Dundee
   The Launch of the SS *Terra Nova* and her Early Years

2 Across the Atlantic and New Owners — 43
   The Ambitions and Achievements of Benjamin Bowring
      and his Family
   The Founding of a Shipping and Trading Company
   '*Terra Australis Incognita*'

3 From the North Atlantic to the Antarctic — 55
   Relief Ship for the British National Antarctic Expedition, 1901
   Under Dundee Command Again and First Mission to Antarctica

4 From the Antarctic to the Arctic — 65
   Under US Ownership and Norwegian Command
   Rescue Mission to an Arctic Archipelago
   Return from a Successful Arctic Mission
   Return to Newfoundland for Sealing Duties

5 Her Name Will be Remembered Forever — 83
   Expedition Ship for the British Antarctic Expedition, 1910
   Fitting Out in West India Dock, London
   Preparations and the British Departure from Cardiff
   Around the World to Lyttelton, New Zealand

6  Into the Southern Ocean   123
    Final Preparation and Departure from Port Chalmers,
        New Zealand
    A Storm in the 'Furious Fifties'
    Through the Pack Ice and into the Ross Sea
    Arrival at McMurdo Sound and a Change of Command

7  First Role Complete   153
    Expedition Base Established, Scientific Parties Deployed
    A Surprise in the Bay of Whales
    Scientific Parties Landed and the Return to New Zealand

8  New Zealand Refit and Hydrographic Surveys   163
    Summary of Expedition Relief Voyages
    First Return Voyage to New Zealand – 'The Pumps Again!'
    Winter Cruise and Survey Work

9  Return to Antarctica – First Relief Voyage   171
    McMurdo Sound: Relief and Attempted Relief of
        Scientific Parties
    Second Return Voyage to New Zealand and More
        Survey Duties
    Another Refit in New Zealand and a Tragedy at Admiralty Bay

10  To Antarctica – The Final Relief   179
    Last Passage to Antarctica
    Cape Evans and the Tragic News
    Last Departure for New Zealand and Goodbye to Antarctica

11  The Voyage Home   191
    The Passage to Britain
    Return to Expedition Home Port
    Summary of British Antarctic Expedition Programme,
        1910–13
    Report on the Biological Work Carried Out Aboard
        by Dennis G. Lillie, MA
    Extract from Report: 'Outfit and Preparation'
        by Commander E.R.G.R. Evans

12 Newfoundland and Sealing    213
    Thirty Years with the 'Wooden-Walls'
    Rendering Assistance at a Maritime Disaster
    Portrait of a Legendary Sealing Master, Captain Abram
        Kean OBE
    A First-Hand Experience of the 'Greatest Hunt in the World'

13 Chartered for War Duties    249
    Refit and Role as a Coastal Trader
    On Charter During Wartime
    Ice Damage to the Stern: The Last Voyage and an SOS Call
    Memories and Recollections
    Research and More Recollections

14 The Finding of the Wreck    271

Epilogue    277
Appendices
  A    Ships Built at Dundee by Alexander Stephen & Sons 1844–93    281
  B    Description and Specifications of SS *Terra Nova*    284
  C    Extract from the Log of USCGC *Atak*    298
  D    List of Captains Who Commanded the SS T*erra Nova*,
        1884–1943    300
  E    SS *Terra Nova* Crew List, Antarctic Relief Voyage, 1903–04    307
  F    SS *Terra Nova* Crew List, Arctic Relief Voyage, 1905    308
  G    SS *Terra Nova* Crew List and Shore Parties, British Antarctic
        Expedition, 1910–13    309
  H    Summarised Directory of the Arctic Region and Antarctic
        Continent    312
  I    Alexander Stephen & Sons Ltd, 1750–1970: The Family
        Line of Shipbuilders and a Brief History of the Company    317
  J    Bowring Brothers Ltd, Profit & Loss Accounts Balance
        Sheet, 1943    320
  K    Some Sealing Phrases and Expressions    324
  L    Miscellaneous List of Whalers and Sealers Launched
        and Their Fates    325

References    329
Bibliography    335
Index    338

# Acknowledgements

It would not have been possible to write this story without the records of those who had sailed in the SS *Terra Nova* and my first acknowledgement must be to the men of yesteryear who kept diaries, wrote letters home or recorded in any way the happenings of everyday life aboard ship and the many events, work and expeditions with which they were involved. There will be many stories which still lay undiscovered.

In addition to available publications, it has been necessary over many years to accumulate material from the archives of various institutions and I am grateful for the assistance given by the Scott Polar Research Institute, University of Cambridge; the Norsk Polarinstitutt, Tromsø; the Memorial University of Newfoundland; the Provincial Archives of Newfoundland and Labrador; the Department of Defense Naval Historical Center and the National Archives and Records Administration, Washington DC, USA; the Maritime Archives of the National Museum of Wales; the National Museums of Scotland; the University of Glasgow Archives; the National Maritime Museum, Greenwich; the Oates Museum and Gilbert White's House, Selborne; the British Antarctic Survey and the Archives of the Captain Scott Society.

I am grateful for the assistance of the Guildhall Library, City of London; the City of Cardiff Central Library; the City of Dundee Central Library and McManus Museum & Galleries; the Britannia Royal Naval College Library, Dartmouth; the Maritime Museum of the Atlantic, Halifax, Nova Scotia; the Museums of Canterbury, Marlborough, Nelson and Port Chalmers, New Zealand; HM Hydrographer of the Navy, Hydrographic Office, Ministry of Defence, Taunton; the UK Antarctic Heritage Trust and the Dundee Heritage Trust.

My thanks also to publishers Tundra Books Inc., Montreal and Toronto; St Martin's Press, New York; Flanker Press Ltd, St John's and Mr Clyde Rose of Breakwater Books Ltd, St John's, Newfoundland.

During preparation for this project, I turned many times with queries in the UK to Robert K. Headland and William Mills of SPRI, Sir Vivian Fuchs FRS, Dr Anthony M. Johnson, Rear Admiral John Myres CB, David Yelverton FRGS, Judy Skelton, Tim Hughes of Kew, Dr David Jenkins, Ann Savours Shirley, Dr John Heap CMG, David S. Henderson, Angus Erskine, Herbert Dartnell, Norman Watson, Mr Peter Bowring CBE and Mr Antony Bowring.

My appreciation also to Professor Julian Dowdeswell, Director of SPRI; James Campbell; John A. Davies; Julian D.O. Salisbury; Matthew G.A. Salisbury; Peter W. James; Brian Thorpe; John E. Davies; John S. Cosslett; Richard H. James; Alf H.K Thomas; Peter L. Jones; John H.L. Rowlands; Basil J. Watkins; John M.D. Curteis; Owen Willmer; Paul A. Davies; Derek H. Moss and David Scheeres. My thanks to Gary C. Gregor, Philip Chatfield and to H. Stanley Richards, past curator of the City of Swansea Museum.

Descendants of men who served in the ship have also answered my queries and given their assistance. The Hon. Edward Broke Evans allowed me unrestricted access to his family archives and let me use passages from his father's book, *South with Scott* by Admiral E.R.G.R. Evans. My thanks to Angela Mathias and Hugh Turner, who allowed me to quote from the classic work *The Worst Journey in the World* by Apsley Cherry-Garrard. My thanks also to Dr David M. Wilson FZS for permission to quote from the *Terra Nova* diary of Dr Edward A. Wilson. and to Jo Laurie and John J. Ramwell, descendants of 'Birdie' Bowers, who allowed me to quote from his letters.

Tony & Moira Pennell and Christopher Lee Pennell provided family details of Commander Harry Pennell, and Ellen-Johanne McGhie, daughter of Tryggve Gran, allowed me to quote from his diary, *The Norwegian with Scott*. My thanks to Jack S. Williamson and John Evans of Swansea; Eva Skelton Tacey, daughter of James Skelton AB; also Jean Scholar, granddaughter of PO Parsons.

I am also grateful for the interest and support of Professor Robert Swan and Sir Ranulph Fiennes, and for material provided by Mary Cleveland and Sandra Kyne of St Mary's, Scilly Islands. Also, the reminiscences and material provided by Mrs June Siveyer Watson of Ashburton, Devon. My thanks to Mr Arthur Thomas, Group Scout Leader of the 4th Cardiff (St Andrews) Scout Group for allowing me access to their archives, and to the late Lieutenant Folke E. Swenson USCG and his son, Gus, for their support and interest in the early stages of the project.

## Acknowledgements

Thank you to Ray Harvey (Design), Department of Maritime Studies, Southampton Institute University College and Mr & Mrs Colin Mudie of Lymington; also to the Royal Dart Yacht Club, Kingswear, Devon. I am grateful to the Commodore, Royal Navy Trophy Centre, HMS *Nelson*, Portsmouth and to Mr Geoffrey Buck, the Royal Navy Trophy Officer.

My thanks to Mr Christopher Grant of Forfar, who was able to provide me with family details and photographs of the Stephen family including his great-grandmother, Alice Stephen, who launched *Terra Nova*.

My appreciation to Mr A.M.M. 'Sandy' Stephen, the last managing director of the famous Scottish shipbuilding company, Alexander Stephen & Sons Ltd. He sent me the complete family tree of the Stephen family dating back before the founding of the company. He and his fellow directors were to see the last days of shipbuilding at Linthouse on the River Clyde with an amalgamation of companies and then the closure of Upper Clyde Shipbuilders Ltd in 1970.

A number of people responded to an article on my behalf which appeared in *The Courier* and wrote to me from Scotland with details of family connections and folklore stories; my thanks to them.

Mr Francis Gloak of Perth wrote that his grandfather, David Gloak was a coppersmith at Stephen's yard. He worked on *Terra Nova* and other ships installing boilers and pipework. Mr Roger Buist of West Ferry, Dundee, sent me Dundee whaling songs and wrote that his grandfather William Rodger Salmond, a ship's plater, worked for Gourlay Brothers and at Stephen's Panmure yard. Mr Salmond held a unique form of 'passport' embossed with the Dundee Coat of Arms, signed in those days by the Lord Provost of the City which enabled the holder to travel and gain work in European shipyards.

Mr James M. Robertson of Dundee recalled that his grandfather George Robinson, born 1853, was a cook on *Terra Nova* during a sealing voyage and that the crew used to drink in the town's Arctic Bar, New Inn Entry. Holly Lindsay of Ceres by Cupar, in Fife, wrote that her mother remembered *Terra Nova* anchored off Tayport and connections with Captain Harry McKay and his family; Mr John Aitken of Montrose sent me a file of past local maritime events and Mr John Alexander Smith of Montrose wrote that his uncle, Robert Smith, born 1897, was employed by Alexander Stephen & Sons for thirty-two years at their Linthouse yard on the Clyde.

My thanks must go to Arthur R. Railton, of Martha's Vineyard Historical Society, Edgartown, Massachusetts, USA; David Harrowfield, Christchurch, and Jim Wilson, Port Chalmers Museum, New Zealand; Clive Goodenough of Melbourne, Australia, formerly of Christchurch, New Zealand; Dr Albert E. Jones, Varrick Cox, Dr Shannon Ryan, Heather Wareham, Jack and Rosalind

Power of Fort Amherst and Captain Thomas H. Goodyear of St John's, Newfoundland; John and Gwen Reed, Bridgewater, Nova Scotia; Captain Ulf S. Snarby, Liverpool, Nova Scotia; Heddi Siebel, Cambridge, Massachusetts, USA; Kjell G. Kjaer, Torsvaag, Vanna Island, Nórway and to Assa Sonberg Kendall, formerly of Oslo, who kindly translated Norwegian correspondence and data for me.

Finally, my thanks to Ann Griffiths who acted as proofreader to check my presentation, and her husband, John C. Griffiths (FIDS Halley Base, 1960–63), who was well qualified to make helpful observations, and to Bob Headland, of the Scott Polar Research Institute who scanned my historical references, most appropriately, during a voyage in an icebreaker while on passage through the Arctic Ocean and the Northwest Passage.

Michael C. Tarver, Hillhead, Brixham, Devon, UK<br>March 2020

## Author's Note

In putting together this story, I have sought to take the reader back to the mood of the times and have tried to avoid straying from the main theme. This was not easy, for to write about just one ship without mentioning other vessels and events of those days would not have provided a sufficiently comprehensive historical picture nor have been of engaging interest. This is essentially a story of the men who sailed in the ship. I have used their own words in many paragraphs and for the sake of continuity, where I have found it fitting, I have linked relevant parts of their stories using my own words.

Readers might find it helpful to copy some of the maps and plans included so that they can follow the story more easily and track the many place names mentioned, particularly following the movements of SS *Terra Nova* when in McMurdo Sound, Antarctica.

Before publication, I sought the help of the media in Scotland and Newfoundland. I visited Dundee and St John's to seek facts and to give an opportunity for anybody to make contact with me who might have useful information both about *Terra Nova* and the Dundee shipyard of Alexander Stephen & Sons Ltd. I am most grateful to those who wrote to me. I was particularly pleased to hear from descendants of the Stephen family, whose forbears built the ship, and members of the Bowring family whose company were her owners for forty-five years.

I am sure there is much about the ship that I have not discovered. It is inevitable that more information will come to light following publication, and this I welcome. Using many original sources, every effort has been made to keep to historical accuracy. I welcome any further details which interested readers may offer. To assist research in the future, they can communicate with me by writing to my publishers or to the Scott Polar Research Institute, University of Cambridge, in whose archives the SS *Terra Nova* file is to be found.

In linking the various sources of the story, the views expressed are mine alone. They are my thoughts after I have studied the heroic age of polar exploration for twenty years, including undertaking a voyage to the Ross Sea region of the Antarctic continent in an icebreaker. I visited Dundee to see what is left of the Panmure shipyard where *Terra Nova* was built and, to make my story complete, I felt compelled to visit St John's, Newfoundland to get to know the port from which *Terra Nova* had operated. Here I met many interesting and helpful people whose warmth and hospitality helped me to feel part of the history of this famous fishing and sealing port.

I acknowledge with particular gratitude the kindness of Her Royal Highness the Princess Royal in writing the foreword to this story.

Michael C. Tarver

# Prologue

The selection of the whaler and sealer SS *Terra Nova* for polar exploration and relief duties on three occasions amounted to a total engagement of six years in the ship's working life of sixty years. For the majority of her years she performed mercantile duties, initially as a whaler from Dundee and then in the sealing trade from St John's, Newfoundland. In her last years she was chartered as a wartime supply vessel and it was during this role that she foundered.

The acquisition of *Terra Nova* as expedition ship for the British Antarctic Expedition in 1910 is what brings her to notice as a famous ship and gives her a place in history. As a relief ship, seconded from her sealing duties in the North Atlantic and Arctic regions, she had proved herself a formidable vessel suitable for use in ice. It was her selection as the expedition ship for Captain Robert Falcon Scott's last expedition to the Antarctic that secured for her the fame with which she will always be associated and assured her a place in the history of the sea and polar exploration.

There is a long story to tell of her service in the North Atlantic seal fishery and earlier whale trade. *Terra Nova* was a familiar sight at her home port of Dundee, Scotland, and subsequently in St John's, Newfoundland, where for so long she had a special place in the hearts of two seagoing generations on both sides of the Atlantic. This is the story of the ship and her many roles.

There are many written accounts of exploration in the polar regions, attempts to cross the Arctic Ocean toward the North Pole and penetration of the Antarctic continent toward the South Pole. Accounts have been written by many of the explorers themselves, with later contributions by other authors who seek to analyse the stories and tell of the achievements, failures and incredible hardships involved with polar exploration. All make exciting reading for armchair explorers and lovers of the sea.

An account of life aboard *Terra Nova* which describes five weeks of sealing from Newfoundland was written in 1922. It is fortunate that we can include excerpts from that fascinating period, together with some records of the company that owned *Terra Nova* for forty-five years. Her principal home port was St John's and her owners, Bowring Brothers Ltd, operated with its parent company, C.T. Bowring & Co. Ltd, of Liverpool and London. They acquired *Terra Nova* from her owners at the port of Dundee where she was built and operated for the first fourteen years of her life.

Between 1895 and 1922 is the period generally referred to as the 'heroic age' of Antarctic exploration, where territorial gains and scientific discoveries were made by expeditions attempting to extend the boundaries of man's knowledge of the continent and to pursue the quest to reach the South Pole. This coincided with attempts to cross the Arctic Ocean and to be first to reach the North Pole.

The leaders of these polar expeditions remain household names. The lives and careers of the many men that accompanied them reveal them to be equally brave and fascinating, but little has been written about the ships that took them there. These were wooden ships, being used apparently out of their time as the world advanced from the steam age into the twentieth century. Composite and iron ships had been built since the mid-nineteenth century, but these early designs proved totally unsuitable because condensation made life unbearable for the crews in the cold regions and the iron hulls were easily damaged by ice.

The Royal Research Ship *Discovery*, built of wood, was used for two scientific expeditions to the Antarctic regions by Sir Douglas Mawson as late as the 1930s, and another wooden ship the SS *Bear of Oakland* was used by Admiral Richard Byrd for Antarctic expeditions up to 1941.

Welded ice-strengthened steel construction gradually improved, and nowadays powerful icebreakers have long taken the place of those early wooden vessels which had their bows sheathed in metal plates. Shipbuilding technology developed further during the Second World War, when welded vessels known as 'Liberty' ships were built.

Nowadays, ships with purpose-designed hulls for work in the ice fields are able to go virtually anywhere during the summer in the polar regions. During the heroic age, the only suitable ships were those with wooden hulls. Fitted with iron-sheathed bows, they could enter the ice and the hulls were able to flex with pressure from the ice to reasonable limits.

SS *Terra Nova* began her life as a steam-assisted barque built in the late nineteenth century at a time when whale and seal products were still in demand. However, in that declining market she carried on through a transitional stage in history and gave service for a total of sixty years from her launch in 1884 to her end in 1943. *Terra Nova* was one of many wooden ships from that era that had been suitable for exploration of the polar regions. Expeditions of the time show a list of British and Norwegian wooden ships in use during the heroic age, many having careers of different lifespans and mixed fortunes.

Records show that most of the nineteenth-century British ships built as whalers and sealers and used in the polar regions came from the Dundee yard of Alexander Stephen & Sons (see Appendix A). Some similarly designed vessels were built in other British yards.

Norwegian shipyards also built whalers and sealers and explorers of many nations were pursuing interests in the polar regions, among them France, Germany, Russia, Norway, Sweden and Japan. There are many stories on record, including the losses of ships.

There is a never-ending fascination with the stories of these expeditions, the leaders, their men and their adventures. But what happened to the ships of the heroic age, each one an essential part of an expedition; each ship with its own personality and characteristics and as much a part of the expedition as the men themselves? What became of these vessels that served them so well?

*Southern Cross*, formally a Norwegian whaler named *Pollux*, was built in Arendal, Norway in 1886. She was the expedition ship used by Carsten Borchgrevink and the British Antarctic Expedition, 1898–1900. Under the command of Captain George Clark, *Southern Cross* went down in 1914 off Newfoundland. She was last seen off Cape Pine when returning from sealing in the Gulf of St Lawrence. All 177 men aboard were lost. A single lifebelt, months later, was picked up off the coast of Ireland, but was never positively identified as coming from that ship.

*Scotia* was used by Dr William Spiers Bruce for the Scottish National Antarctic Expedition, 1902–04. Formerly the Norwegian whaler *Heckla* and built in 1872, she was chartered by the Board of Trade to monitor ice conditions in the North Atlantic following the *Titanic* disaster in 1912. Efforts were made to conserve her but early in the First World War she was sold and ended her days as a collier operating out of Cardiff. On 18 January 1916, loaded with coal, on passage to a French port, she caught fire just after leaving Cardiff and was beached by tugs at Sully Island in the Bristol Channel. She became a total loss, although fortunately without loss of life.

The sealer and whaler *Nimrod* came from the yard of Alexander Stephen & Sons in 1866 and was chosen for the British Antarctic Expedition, 1907–09 led by Ernest Shackleton. In 1911, she took an expedition to the Yenisei River, Siberia. During the First World War she served as a coastal collier. She was wrecked in the North Sea on the Barber Sands off Caister, Yarmouth, on 29 January 1919. Her twelve-man crew launched the ship's lifeboat, but it then capsized. There were only two survivors.

Sir Ernest Shackleton's *Endurance*, formerly the Norwegian barque *Polaris*, built in 1913, was watched by her crew as she was crushed in the ice and lost in the Weddell Sea in 1915, after the Imperial Trans-Antarctic Expedition had been unable to set up its first base.

The ship always described as the 'gallant little *Morning*', formerly the Norwegian whaler *Morgen*, on two occasions relieved *Discovery* and the British National Antarctic Expedition of 1901–04 led by Scott. *Morning* was lost in the North Atlantic in December 1915, while on convoy duty carrying munitions to Russia.

*Aurora*, 580 tons, was another whaler and sealer from the yard of Alexander Stephen & Sons, built in 1876. She had a truly remarkable series of adventures in the Arctic even before being used on two Antarctic expeditions. In the Antarctic she was used by Sir Douglas Mawson in 1911–14 and then on behalf of Sir Ernest Shackleton in 1914–17 as the second vessel and part of his Imperial Trans-Antarctic Expedition. She gave splendid service during the heroic age and her incredible story, both in the Arctic and Antarctic, deserves a separate study and ought to be fully researched and recorded. The dramatic story of the Ross Sea party of Shackleton's Trans-Antarctic Expedition 1914–17, with its tragedies, and the account of *Aurora's* drift with all its consequences, is told by Richard McElrea & David Harrowfield in *Polar Castaways*.

*Aurora* was lost on a commercial voyage from Australia to South America. She left Newcastle, New South Wales on 20 June 1917 with a cargo of coal but failed to arrive at Iquique, in Chile. Aboard her was James Paton, a seaman with a proud record of service to the heroic age. He had been in *Morning* for the relief of *Discovery*; with Shackleton on the *Nimrod* expedition and was in the crew of *Terra Nova* with Scott.

Later, he went in *Aurora* with Shackleton's Trans-Antarctic Expedition and was in the crew during *Aurora*'s drift. He stayed to work on board, but on her last passage in the Pacific she foundered and was never heard of again. *Aurora* was posted missing at Lloyds on 2 January 1918. The *Cardiff & South Wales Journal of Commerce* quoted a Lloyds report of the loss of *Aurora* on 29 December 1917. The journal lists the names of merchant ships posted missing at Lloyds during wartime and also makes mention of warships lost, but for obvious reasons not named. *Aurora* was listed as 580 tons, O.N. 75106 and her master was Captain Reeves.

On hearing of the loss, the explorer Sir Douglas Mawson, who had sold *Aurora* to Shackleton in 1914, is quoted as saying, 'She was the finest ship on the seas for polar exploration', and he believed that her loss was due to enemy action. However, Morton Moyes, a member of Mawson's expedition who had been in *Aurora* for the relief of the Ross Sea party, suggested that she had been strained by many years of working in the ice and had been overwhelmed by heavy seas. Only a bottle and a lifebuoy were found, washed up on an Australian beach.

During 1944 and 1945, at the time of the Second World War, the British Government consolidated British bases in the Antarctic in what was known as Operation Tabarin. The Norwegian-built wooden sealer, SS *Eagle* was sent from Newfoundland and assisted HMS *William Scoresby* and SS *Fitzroy* in setting up bases on the Antarctic peninsula. *Eagle* was scuttled by her owners after fifty years of service.

As late as 1947, when Sir Vivian Fuchs was appointed to the Falkland Islands Dependency Survey as Field Commander of British bases in the Antarctic, his expedition sailed south in *John Biscoe*. She was the former US boom defence ship *AN-76*, which had served during the Second World War as HMS *Pretext*. She was built of pitch pine and when ice damage to the hull was made good in the Falkland Islands she was fitted with steel plates fastened to her wooden hull with bolts. Sir Vivian recalled to the author that the temporary repairs were regarded with some concern as to how they would withstand the ice conditions.

Two ships from the heroic age still survive, the *Fram* and the *Discovery*. Roald Amundsen's ship, *Fram* was built in 1892 by Colin Archer at Larvik, Norway, for Arctic exploration by Fridtjof Nansen and is now preserved as a national monument in Oslo.

*Discovery* was launched in 1901 by the Dundee Shipbuilders Company, successors to Alexander Stephen & Sons. She was built specially to order for the Royal Society and the Royal Geographical Society, promoters of the British National Antarctic Expedition, 1901–04 led by Commander Robert Falcon Scott RN. After her return from the Antarctic in 1904, *Discovery* was sold by the Admiralty, who had acquired ownership, to the Hudson Bay Company to operate as a cargo ship. Her story is told in *The Voyages of the Discovery* by Ann Savours. RRS *Discovery* is now an exhibition ship and afloat today at Dundee, a tribute to those well-proven ships able to work polar regions and to the men who built and sailed them.

Ask anybody to name Captain Scott's ship and most will reply, '*Discovery*!'. However, in 1909 *Terra Nova* was bought by the Expedition Committee for Scott's second expedition to the Antarctic in 1910. The more authentic answer to the question is actually, *Terra Nova*. Scott's name as managing owner is recorded on the Board of Trade Agreement and Account of Crew.

The career of these historic ships varied according to fortune, but no ship of the Antarctic heroic age could match the full and active sixty-year career of the SS *Terra Nova*, the second largest and most powerful whaler and sealer to be built by Alexander Stephen & Sons.

After her launch in 1884 and into the declining years of the whaling trade, she was twice a relief ship to polar expeditions, first in the Antarctic and then the Arctic. Her reputation established, she was then thrust into her most famous role. Having been chosen as a polar exploration ship, she went back to the Antarctic. Afterwards she returned to sealing in the North Atlantic and finally she was chartered for wartime supply duties, during which she foundered. Thus, six of her sixty years were spent with polar expeditions, the remainder as a whaler and sealer and finally as a cargo vessel.

It is of interest that when making comparisons with other vessels built by Alexander Stephen & Sons at Dundee there was one sealer larger than *Terra Nova*, and another which sailed the seas for eighty-nine years. Both of these are worthy of special mention.

The largest wooden vessel to be built at Stephen's yard at Dundee was the sealer, *Arctic II*, completed in 1875. She was built to replace another vessel of that name which had been lost in 1874. A steam assisted barque, her dimensions were 200.6ft x 31.6ft x 19.9ft and 828 tons (Builders Old Tonnage gross). She had an early end, being lost in the ice in 1887.

The longest service record is held by the *Bear*, a steam-assisted barque of 689 tons launched in 1874. She served for eighty-nine years until she was lost in 1963 while under tow. Mainly US owned, she operated in the Arctic regions and also in the Antarctic but in a period after the heroic age.

In the Arctic she had taken part in the rescue of abandoned whalers and sealers left on the ice after the loss of their ships and had been part of the mission with the Dundee whalers *Thetis* and *Aurora* to rescue the US Arctic explorer, Adolphus Greely, in 1883. *Bear* was involved in events at the Alaskan north-western town of Nome where there was some trouble at the time of the gold rush in 1899. During the First World War she was a revenue cutter at San Diego and later a 'rum chaser' during the prohibition. In the early 1930s she was a museum ship in the City of Oakland, near San Francisco. Here, she was put up for auction and bought by Admiral Richard Byrd, US Navy for an Antarctic expedition and renamed, *Bear of Oakland*.

In 1935, on return from the Antarctic, *Bear of Oakland* was sold by Admiral Byrd to the US Navy for $1 and became the USS *Bear*, taking part in his third expedition to the Antarctic from 1939 to 1941. In 1948 *Bear* was sold to the Shaw Steamship Company of Dartmouth, Nova Scotia, for sealing. She was sold again in 1963 to go on public display in Philadelphia but while under tow she sank off Nova Scotia. Her bell is displayed in the Explorers' Club, New York, the city to which years before she had safely returned the survivors of the Greely Arctic Expedition.

The varied life of SS *Terra Nova* during her sixty years of whaling, sealing and polar exploration in the heroic age is told using entries from the ship's logs, diaries, journals and letters sent home by those serving aboard her to family and friends, and from books and accounts written by those men. This story is dedicated to these men. Most of what the author has written and linked together belongs to them. Much of it is written in their own words just as it was intended to be read.

# Foreword
## by HRH Princess Anne

BUCKINGHAM PALACE

It is surprising that the story of the SS Terra Nova has not been told before, considering the prominent part the ship played in expeditions during the 'heroic age' of polar exploration. The life-span of the ship covers a fascinating period in history and illustrates the rapid industrial, maritime and social changes that have taken place since the late 19th century. Even the name Terra Nova associates her with the part she played in the industrial development of an island community, Newfoundland, and also the early exploration of the unknown continent of Antarctica.

It is unlikely that any other ship with such an adventurous past has had so much written down about her by the men who actually sailed in her, and as a tribute to their enterprise and exploits many of these accounts have been linked together to tell the story of Terra Nova. Scientific study of the polar regions is now recognised as vital in assisting us to protect our environment and in the study of climate change. It is to the early explorers of the polar regions in ships like the Terra Nova that we owe our gratitude for their courage and discoveries as we attempt to take their work forward.

*Anne*

# Introduction
## by Ranulph Fiennes

This second-edition story of SS *Terra Nova*, a famous Scottish-built ship, brings together so many tales of maritime history, not only of the vessel itself as told by the men who sailed in her, but equally through the development of a flourishing trade in animal oils, combined with the increase in demand for general cargos around the world.

In the nineteenth century the shipbuilding company of Alexander Stephen & Son prospered through the days of sail and steam, while at the same time Benjamin Bowring and his family developed their business in Newfoundland as ship owners, general merchants and insurance brokers. The SS *Terra Nova* was to become owned by the Bowring company in 1898.

This development was to coincide with territorial exploration and the advancement of science in the polar regions, where such ships were known to be ideal for the rigours of ice work. The 'heroic age' of polar exploration saw many wooden ships used by expeditions and, years later, when we set out on the Transglobe Expedition, we were so proud that the Bowring company rekindled its links with both the Arctic and Antarctic by providing us with our ice-strengthened ship MV *Benjamin Bowring*, named after the founder of that historic company.

Mike Tarver's fascinating and detailed history of the SS *Terra Nova* shows just how tough those early days of polar exploration really were.

Ranulph Fiennes
September 2019

# 1

# INTO THE EVENING OF A PASSING AGE

## Introduction

This is the story of a ship – a whaler, sealer and polar exploration ship that sailed the seas for sixty years. From her launch in 1884 she sailed the oceans from the Arctic regions to Antarctica until she foundered in 1943. Once she was launched, her story can only be told by the men who sailed in her – they bonded with the ship in which they placed their trust when they put to sea.

How does one tell the story of a ship? To those who do not know the sea, a ship is an inanimate object but to those who have been carried over the oceans, who have been close to nature in all conditions, who have seen the horizon under a ceiling of stars, felt that feeling of humbleness beneath the power of the universe and experienced the moving awesomeness of the sea which presents a constantly changing picture, a ship seems to have a heart, a soul and a spirit of her own. This also applies to smaller craft on inland waters – whether you have contributed to her construction and equipment or merely been a passenger, you must place your trust in your ship. Ask any old seaman his feelings on seeing his ship go down, perhaps torpedoed in wartime, or ask a captain who takes his ship to be decommissioned. Ask any yachtsman who sells his boat which has given him many years of pleasure. Whatever the vessel, whatever her character, whatever stretch of sea or water is crossed, a special relationship grows between 'man' and ship. Your trust is in the vessel which carries you – and a ship *must* be a 'she'.

It is rare for a ship to sail the seas for sixty years, particularly a ship built of wood. She will have many tales to tell. Life will begin on the drawing board and as her lines take shape, the men who build her begin a relationship themselves as they put together their skills toward the day she is launched. Most ships are built for a specific purpose and are suitable only for that role. Few last for as long as sixty active years. Not so the SS *Terra Nova*, designed and built as a whaler and sealer in the tradition of her time.

Towards the end of an era in nineteenth-century Britain, in a country with an empire built on an economy driven by coal and steam, *Terra Nova* entered into an age of ever-changing technology and still kept sailing, changing roles as the times demanded. Parts of the story of *Terra Nova* have already been told by the men who sailed in her. The author has sought to discover those stories spanning her entire career and bring together the tales of those men of yesteryear.

## The British Whaling Industry and its Ships

Animal oils from whales, seals and fish were much used in the manufacture of foods, fabric products, soap, fertiliser and lighting until the late nineteenth century. From the early nineteenth century, the whaling and sealing fleets of Britain were sent from many ports, principally Hull, Aberdeen, Peterhead, Newcastle and Dundee to Arctic waters in search of their catch.

As iron and composite ships were developed and steam power was introduced, entrepreneurs thought they could increase their catch by sending out ever larger vessels. Crews soon found that working aboard the very early iron ships brought problems in cold regions. The ships were damp, cold and dripped with condensation. Their rivetted, plated iron hulls became damaged in the ice and the engines frequently broke down.

All this was much to the delight of traditional crews in wooden ships, which sometimes had steam engines but were always rigged for sail and could be sailed out of trouble. These crews would often be called on to help other crews in their icebound and damaged ships. Sometimes help couldn't be given or was perhaps even refused. There were accidents and there was loss of life.

One port, above all, which persisted with whaling and sealing by using traditional wooden ships assisted by steam power was the Scottish port of Dundee, which, as records show, had sent out whaling ships since 1750. When the early iron ships failed, the activity at other ports diminished, but Dundee prospered and continued to build ships with wooden hulls.

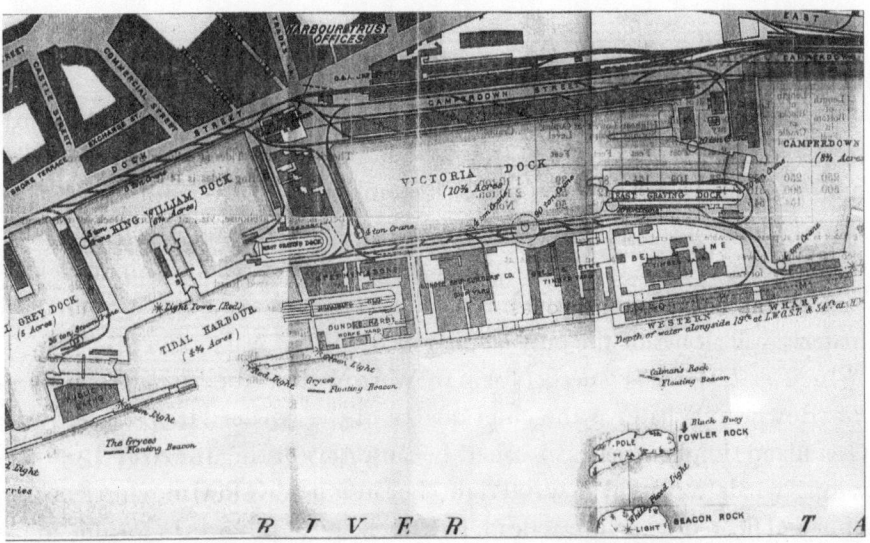

Dundee Docks and tidal harbour. (*British Docks Association Handbook*, 1912)

By the first half of the nineteenth century, with the coming of the railway, Dundee became established as a major engineering centre in Scotland. Shipbuilding and railway industries and jute and hemp importing businesses developed there and the port became the centre of the British whaling industry. With this growth in employment and prosperity, Dundee became important enough to be granted a royal charter as a city by Queen Victoria in 1889. One of the local industrialists, Sir James Caird, a jute and hemp importer, became a great benefactor to the city.

The earlier years of whaling in ships without steam power carried a high cost in loss of human lives and shipping. Tragedies in Arctic regions, particularly the Davis Strait, were commonplace, with many ships failing to return. Whole crews were lost or left abandoned on the ice, either to perish of the cold or, if they were lucky, to be rescued by another ship. This was the way of life in those days. Dundee produced many famous whaling skippers, amongst them Captain William Adams and Captain Charles Yule, who became harbourmaster at Dundee.

The early nineteenth century saw the development of docks along the banks of the River Tay. Between 1815 and 1830 the King William IV Dock and Graving Dock were completed with sea walls and quays forming the Tidal Harbour. Then came a fine custom house in 1843. Later came the Earl Grey Dock, the Victoria Dock and then the Camperdown Dock. Here, at Dundee, a growing and prosperous town with fine maritime facilities and an allied industry, was the centre of Britain's whaling fleet.

By the middle of the nineteenth century, Dundee had grown into the third-largest town in Scotland after Glasgow and Edinburgh. The number of textile mills and the growth of steam power technology increased rapidly, providing employment for the growing population. This led to increasing activity at the docks where exports of the cloth, sack and bag by-products of imported jute and hemp combined with the hustle and bustle of the importing of the raw materials, all alongside the busy landing of whale and seal products.

In *The Dundee Whalers*, Norman Watson describes *Balaena mysticetus* (Bowhead Whale) as the chief quarry of the whalers. It was called the Greenland 'Right' whale, so-called because this distinguished it from the wrong whale to catch! It was 'Right' because it was slow moving, easy to hunt and floated when it was dead. This docile giant possessed a fortune in its cavernous upper jaw, overlapping plates of springy, tough baleen ( sometimes referred to as whalebone), which was valuable for use in ladies' corsets and other flexible products. No other whale had such a mass of baleen and this made the 'Right' whale a prime commercial target. It also had a thick layer of fat, or blubber, a valuable commodity when boiled to oil.

Certainly, if the ships came home with a good catch financial rewards were greater than could be earned ashore, but the risks to human life were enormous. Even the introduction of steam power in the middle of the nineteenth century, while bringing many benefits, still brought extra loss of life as more ambitious attempts were made to bring home the catch. The biggest danger was that the ship might be 'nipped', i.e. squeezed or crushed by moving pack ice. Many were lost in this way, the crews being left on the ice dressed only in what they stood up in and with what possessions could be scrambled from the ship before she disappeared through the ice. Such losses must have been dreadful for the surviving crews and those who rescued them.

As towns and cities grew in the late nineteenth century, demand for heat, light and power came to depend more on coal and gas. In 1885 the Austrian scientist Carl Von Welsbach (1858–1929) invented the gas mantle for lighting. Gasometers sprang up in populated areas, and the demand for animal oils declined and market prices for whale products dropped. At the same time, there was a danger that the 'Right' whale might be hunted to extinction. An era appeared to be ending.

In 1841 Captain James Clark Ross had sailed his ships HMS *Erebus* and HMS *Terror* through the pack ice into the sea round Antarctica, which is now named after him. He had reported great numbers of whales for the taking.

Dundee Seal & Whale Fishing Co., St John's, Newfoundland (est. circa 1870 for co-operation of men and resources). Photo taken around 1880. Back row (left to right): Captain Alexander Graham (*Bloodhound*), Captain McLennan (*Narwhal*), Captain James Bannerman, Captain Charles Dawe (*Aurora*), Hon. Moses Monroe (Company founder), Mr John Pye (shipbuilders, Alexander Stephen & Sons, Dundee), Mr. A.G. Smith (broker). Front row (left to right): Captain John Green (Tug Co.), Captain Charles Yule (*Esquimaux*), Captain William Adams Snr (*Arctic*), Mr Francis Winston (editor of the *Morning Chronicle*). (Courtesy Memorial University of Newfoundland)

Perhaps an answer to the decline of the whaling industry might be found in the southern hemisphere? In hopeful expectation of a sea of plenty, ship owners and shareholders decided to send a fleet south.

The Dundee Antarctic Whaling Expedition of 1892–93 consisted of four ships which sailed to Antarctic waters with the hope of discovering new and plentiful hunting grounds, but the speculators were to be disappointed. Captain Alexander Fairweather in the whaler *Balaena* led the fleet, accompanied by *Polar Star* (Captain James Davidson), *Active* (Captain Thomas Robinson) and *Diana* (Captain Robert Davidson). They sailed via the Falkland Islands to the peninsular region of the Antarctic continent.

Unfortunately, no 'Right' whales were to be found, only an abundance of a species not profitable to pursue. The fleet returned to port some nine months later with a cargo of sealskins and oil, together with some scientific and geographical information, all much to the financial disappointment of those who had funded the project. It is of interest, however, that in probing south, these were the first power-driven ships to have penetrated the Antarctic Circle since the scientific voyage of HMS *Challenger*, when, on 16 February 1874, *Challenger* crossed the Antarctic Circle and stood on for 10 miles before turning north. These events predated the heroic age of Antarctic exploration but might be regarded as a practical beginning.

It is interesting to note that of these four Dundee-based ships that went south on the expedition of 1892, only *Diana*, at 473 tons, was built at Dundee by Alexander Stephen & Sons and launched in 1891 for Job Bros of St John's. *Balaena* was built in Norway, *Active* and *Polar Star* were built at Peterhead. *Diana* is shown bringing in a seal catch at Newfoundland early in 1892 after which she joined the fleet for the Antarctic venture which left Dundee on 6 September 1892.

Today, the city of Dundee, its maritime and associated industrial past now left to historians to record, remains famous for its marmalade and Winston Churchill's favourite fruit cake. As a reminder of nineteenth-century times, Dundee is home to the Royal Research Ship *Discovery*, the last example of a nineteenth-century wooden ship based on the design of a Dundee whaler built in the yard of her forerunners. Her keel was laid down on 16 March 1900 and she was launched on 21 March 1901 specially for use in polar exploration.

She is now restored and preserved as an exhibition ship. Though *Discovery* was never a whaler or a sealer, the visitor strolling round her decks can easily imagine all the excitement and tension of those days. When taking dinner in her splendid wood-panelled and elegant wardroom, it brings to mind all the nostalgia of being part of another age when the furtherance of polar exploration to extend man's knowledge of the globe was so geographically and scientifically important.

The *Discovery* is a symbolic reminder of those famous years when, despite all the hardships and losses of life in the whaling industry, the city of Dundee grew and developed as a world centre for whaling, together with the import and manufacture of jute and hemp products. Here, too, evolved the skill of constructing wooden ships. Dundee people proudly built the western world's finest whalers and sailing ships. Towards the end of the era, SS *Terra Nova* was the second largest, most powerful (and, indeed, the last) whaler to be built for active duty in those days.

## The Shipyard of Alexander Stephen & Sons, Dundee

The city of Dundee is set on the north bank of the River Tay, and downriver from its famous railway bridge was once situated the entrance to a tidal harbour and the city docks. Industrial archaeology reveals some of what remains of the nineteenth- and early twentieth-century shipyards. A slipway at what was once known as Stephen's Yard, the Panmure Shipyard, Marine Parade, still remains almost 'spooky' within a restructured industrial area. Here can be found traces of Alexander Stephen & Sons' Dundee yard. Some of the old dockyard buildings which would have housed offices and ancillary services to the industry are still to be seen. Here would have stood the sheds from which the smell of boiling whale blubber would have wafted across the town.

In those days we would have entered the Tidal Harbour from the River Tay and turned to starboard (that is, right). There, we would have seen the shipyard of Alexander Stephen & Sons. The yard extended east beyond the buildings, and tall sheds held launching slips which ran directly into the river. This area later became a timber yard, now lost. Much alteration has taken place. Within the past twenty-five years the building of a new road bridge over the river and a new road network embracing the old city docks has changed the face of the waterfront. The plans included a dock for the RRS *Discovery* and a visitors' exhibition centre.

The old tidal harbour has been filled in, leaving what remains of Stephen's shipyard to the imagination. The Stephen family began building small boats for fishermen at Burghead on the Moray Firth in 1750 and afterwards at Aberdeen. The railway had reached Dundee in 1838, and at about the same time came the introduction of the screw propeller and the development of steam power in ships.

By 1843, the family firm became established at what became known as the Panmure Shipyard, Marine Parade, Dundee. There were some setbacks. In 1868 the Dundee yard, with two unnamed ships on the stocks, was totally destroyed by a fire which started near the sawmill. One ship was a composite and the other a whaler, both well advanced and almost ready for launching. Two hundred employees were thrown out of work. However, perseverance triumphed in the face of adversity. Within a year the shipyard had been rebuilt and a ship was ready for launching.

The performance of Dundee whalers caught the eye of the sealing industry in Newfoundland and ships were built for a number of companies operating at the ports of St John's and Harbour Grace. From 1851, Alexander Stephen & Sons expanded the Glasgow business on the River Clyde, first at Kelvinhaugh

Tidal harbour entrance and Panmure Shipyard. (Courtesy Dundee Museums and McManus Art Galleries)

Tidal harbour entrance and Panmure Shipyard. (Courtesy Dundee Museums and McManus Art Galleries)

and afterwards at the Linthouse Yard, building large steam ships. The family business passed down the line and continued well into the twentieth century.

Alexander Stephen (1795–1875) was a national figure whose career as a Scottish shipbuilder marked him out in stature and accomplishment as one of the great Scottish industrialists of the nineteenth century, with his sons alongside him. Their company eventually became the second largest of the Scottish shipbuilders on the Clyde. Their business eventually became part of Upper Clyde Shipbuilders Ltd, which went into voluntary liquidation in the 1970s, part of the sad decline of British shipbuilding in the twentieth century. (For the company history see Appendix I.)

Alexander Stephen (1795–1875). (Courtesy Mr A.M.M. Stephen)

## The Launch of the SS Terra Nova and her Early Years

In the list of ninety-seven ships built at the Dundee yard of Alexander Stephen between 1844 and 1893, *Terra Nova* is listed as No. 84 (see Appendix A). That list of whalers and sealers contains the names of many famous ships such as *Aurora, Arctic, Bear, Esquimaux, Nimrod, Thetis, Ranger, Neptune* and *Diana*, whose names crop up in connection with *Terra Nova* in Dundee and Newfoundland and some in association with polar exploration in addition to whaling and sealing. Many other ships on the list were lost before *Terra Nova* was built.

*Terra Nova* is shown as a sealer 187ft long x 31ft beam x 19ft draught, of 744 tons (gross in Builders' Old Tonnage). She was laid down by Alexander Stephen's eldest son, William, to the account of the company. William is shown as owning all sixty-four shares. The ship was launched as a steam-assisted barque on 29 December 1884. Her hull is described as having frames of oak laid on a keel of rock elm and clad with pitch pine and elm. (A more detailed description of her construction is given in the Appendix B.) Her steam engines were built by Gourlay Brothers of Dundee and were two-cylinder inverted compound steam engines of 120hp (combined) with 27in and 54in diameter cylinders, giving a 2ft 9in stroke to a 10½in shaft driving a four-bladed cast-steel propeller.

Gourlay Brothers was a Dundee engineering company which had produced agricultural machinery from the early nineteenth century and then turned to developing marine steam engines. As the Stephen's company activities on the Clyde developed, Ebenezer Kemp, engineering manager at Gourlay Brothers, took his talents from Dundee to the Clyde and entered into partnership with Alexander Stephen & Sons to develop and build engines for larger ships.

The engine details of *Terra Nova* recorded on registration documents at the time of launch show 120hp, though 140hp is shown as her power in a following chapter. She was rigged as a barque, i.e. fore and main mast square-rigged, mizzen mast rigged fore and aft. Her official number was ON 89090. If there were original plans for her building, they have not come to light. Such were the skills of shipwrights in those days that individual plans for each ship were not always drawn up.

William Stephen (1826–93). (Courtesy Mr Christopher Grant)

*Terra Nova* was launched in 1884 by William Stephen's daughter, Alice. The *Dundee Advertiser* reported on 30 December 1884:

> Yesterday the whaler Terra Nova was launched from the shipbuilding yard of Messrs Alexander Stephen and Sons. This vessel has been built to replace the Thetis which was sold to the American Government last year for the Greely Relief Expedition. The Terra Nova has been specially built for Arctic Navigation. It is calculated she will carry about 40,000 seals or 260 tons of oil. She will be employed in the Newfoundland and Davis Straits whale industry and be commanded by Captain Fairweather.

So began the life of a ship which, when her intended career as a whaler and sealer is coupled with her more famous role as a polar exploration ship, became the longest serving and most travelled expedition vessel to survive the Antarctic heroic age. She had sixty years' distinguished service, and it could be said that her excellent service has never received the true recognition it deserved.

*Terra Nova* was based at Dundee for her first fourteen years. She was under the command of Captain Alexander Fairweather from 1884 to 1888. In 1889, and thereafter to 1893, the year which included Captain W. Archer also of Dundee, she was under the command of the Newfoundlander, Captain Charles Dawe.

Alexander Fairweather's brother James had earlier commanded the whaler *Aurora* for many years. During those years, *Terra Nova* and other Dundee whalers would sail to the sealing grounds with large crews taken on at Dundee and in Newfoundland. Here, they would join the fleet from St John's and take part in the seal hunt in the spring of each year. Newfoundland law prescribed that this would not commence before 10 March.

Early in the 1870s the Stephen company had leased ground on the south side of St John's harbour where it built a yard with essential plant for processing seals. This meant a saving in time as, during the sealing season, ships could discharge their cargo on that side of the Atlantic and return to the sealing grounds rather than sailing back to Dundee.

Alice Stephen (1854–1916), who launched *Terra Nova*, daughter of William Stephen. (Courtesy Mr Christopher Grant)

Poor catches and the loss of a number of ships in the late 1880s began to signal that the end of the whaling era might be in sight. In *The Dundee Whalers*, Norman Watson tells us that although this decade began with optimism, with the port operating seventeen ships, the price fetched by animal oil was dropping and in 1885 two ships were lost. The following year was disastrous, with a further four ships squeezed by ice and lost and the catch brought in by the remainder very poor.

In 1887, ten ships sailed to the ice, but that year saw the loss in the Cumberland Gulf of Dundee's largest whaler, *Arctic II*. This was a severe loss of pride to the fleet. In the same year *Terra Nova*, commanded by Captain Alexander Fairweather, brought in 26,234 seals and a whale, but there was a tragedy aboard when a most respected crewman of the town, James Cummings, aged 76, fell to his death through thin ice. He had been at sea since he was 16 and had made fifty-seven trips to the Arctic.

Earliest available photographs of *Terra Nova*, which show the ship's name on the bow. She was bought by Bowring Brothers Ltd in 1898. The company insignia is shown on the funnel, indicating that the photo above is the earlier one. Note the whale-chaser boats in the davits. (Courtesy of Dundee Museums and Art Galleries and *The Courier*)

In 1890, Captain William Adams, Dundee's most famous whaler, fell ill and died aboard his ship *Maud* on passage home. (He was succeeded in the fleet by his son of the same name.) All these unfortunate events cast a cloud over an already anxious industry.

William Stephen (1826–93), eldest son of Alexander Stephen, had done much to revive the whale trade and had demonstrated his faith in it by building *Terra Nova*. His company also operated the Arctic Tannery at Dundee where they cured and tanned seal skins. William, the last of the family to manage the Dundee yard, died on 7 September 1893 and the yard closed down. He died after a brief illness at his sister's residence at Granton-on-Spey on his way home from a holiday in the Highlands. He had retained ownership in a number of vessels built at his yard which were sold at auction after his death. William had married twice and was the father of twelve children. He was described as a man of untiring energy and perseverance, great organising capacity and unfailing optimism, who infected others with his enthusiasm. He insisted on efficiency and care in construction and maintained a close watch on everything that was done.

Captain Alexander Fairweather (1853–96). (Courtesy *The Dundee Whalers* by Norman Watson)

William's daughter Alice, who had launched *Terra Nova*, married Alexander Gordon Thompson, a Dundee engineer. There are many descendants of the Stephen family, and the family line of shipbuilders can be seen in Appendix I, together with a brief history of the company.

After *Terra Nova*, larger ships both composite and in steel were built, but the end was in sight. The yard at Dundee was proving too small for the type of vessels now in demand and the affairs of the company were sold to the Dundee Shipbuilders' Company Ltd. A number of Stephen's foremen and staff transferred to this company, which was later to build the RRS *Discovery*.

Losses in the volume of ship construction forced the Dundee Shipbuilders' Company into liquidation and it ceased business in 1906, but the Caledon Shipbuilding & Engineering Company, which had unsuccessfully tendered for the *Discovery*, was coming to prominence with the building of steel ships in Dundee. From 1894, all the shipbuilding operations of the Stephen

Rigging plan of *Terra Nova*, stamped 'Admiralty Whitehall'. (Courtesy National Maritime Museum, Greenwich)

company were based on the River Clyde, and continued to be managed by family descendants well into the twentieth century.

Between 1893 and 1895 ownership of *Terra Nova* changed several times. On 28 December 1893, with the sale registered on 5 January 1894, the ship passed into the ownership of William Cox, George Addison Cox, Alexander Henderson and Robert McGavin, appointing as managing owner David Bruce of 3 Royal Exchange Place, Dundee, with Robert McGavin as witness. Andrew Henderson Stephen, Joseph Gibson and George William Lyon Sturrock as executors are shown as the new owners in a variation on 5 January 1894.

After the death of William Cox on 7 September 1894 (the certificate of death is dated 31 January 1895), she is shown as being acquired by George Addison Cox, Alexander Henderson, Robert McGavin and David Bruce Co. of Forfar, merchants, as joint owners. This joint partnership was managed by David Bruce & Co. Ltd and in the years from 1894 to 1897 the ship was commanded by Captain Harry McKay.

In 1894, under Captain McKay's command, records show that *Terra Nova* took five whales and five walruses, making 71 tons of oil, together with 7,232 seals making 70 tons of oil. Company records show that one of the whales captured had embedded in its blubber the head of a harpoon marked 'Jean of Bo'ness' and the date 1854.

Certificate of registry. (Courtesy University of Glasgow Archives)

| | | |
|---|---|---|
| Whether a Sailing or Steam Ship; if Steam, how propelled | Steam; Screw | |
| Where built | Dundee | |
| When built | 1884 | |
| Name of Builder | Alex. Stephen & Son, Dundee | |

| | | Ft. | Tenths |
|---|---|---|---|
| Number of Decks | Two | | |
| Number of Masts | Three | | |
| Rigged | Barque | | |
| Stern | Square | | |
| Build | Carvel | | |
| Galleries | None | | |
| Head | Demi Woman | | |
| Framework | Wood | | |
| Length from Forepart of Stem under the Bowsprit to the Aft side of the Head of the Stern-post | | 187 | — |
| Main Breadth to Outside of Plank | | 31 | — |
| Depth in Hold from Tonnage Deck to Ceiling at Midships | | 19 | — |
| Depth in Hold from Upper Deck to Ceiling at Midships in the case of Ships of Three Decks and upwards | | | |
| Length of Engine Room (if any) | | 31 | 2 |

### PARTICULARS OF ENGINES (if any)

| No. of Engines | Description | Whether British or Foreign made | When made | Name and Address of Maker | Diameter of Cylinders | Length of Stroke | Number of Horses Power (combined) |
|---|---|---|---|---|---|---|---|
| 2 | Inverted Compound | British | 1884 | Gourlay Bros. & Co. Dundee | 27" 54" | 5'3" | 120 |

Master, Alexander Fairweather.
Certificate of Competency: 84,494.

(Signed) E. W. Kyd, Registrar.

### SUMMARY

| Col. 8. Number and Amount of shares of the Transaction, showing how Interest disposed of | Col. 9. Mode of Transaction under which Title acquired | Col. 10. Names of Owners | Col. 11. Mortgages and Certificates of Mortgage | Col. 12. Names of Mortgagees or Attorneys under Certificates of Mortgage | Col. 12. Number of Shares | REMARKS |
|---|---|---|---|---|---|---|
| | 1 | Andrew Henderson, Stephen; Joseph Gibson; and George William Lyon Stewart | Joint Owners | | 64 | Dundee 4th February 1885. William Stephen Junior of Dundee, appointed Managing Owner by letter dated 3rd February 1885, under his own hand. E. W. Kyd Registrar |
| | | Total 28th November 1893 | | | 64 | 12th June 1893. William Stephen is still Managing Owner per reply to Registrar's Enquiry of 27th March 1893 E. H. Hudson Registrar |
| | 2 | William Cox, George Addison, Alexander Henderson, Robert McGavin, David Bruce | Joint Owners | | 64 | David Bruce 93 Royal Terrace Dundee appointed Managing Owner. Advice received 5th January 1894, under the hand of Robert McGavin registered Owner |
| | | Total 5th January 1894 | | | 64 | F. Tolputt Registrar |
| | 3 | George Addison, Alexander Henderson, Robert McGavin, David Bruce M.D. | Joint Owners | | 64 | |
| | | Total 1st February | | | 64 | |

*Lloyds' Register of Ships* for 1860 and 1862 records what is probably the same vessel. *Jean of Bo'ness* was a brig of 163 tons, and her master was Captain J. Harman. She was built in 1803, and was owned by Webster of Faversham (National Museum of Wales archives).

Records show that the owners of *Terra Nova* ceased whaling after 1896, after which she only brought in seals. Financial losses were now being shown by many ships in the industry. The owners of *Terra Nova* were also losing money and faced voluntary liquidation. The sale of the ship was inevitable.

Meanwhile, early in the nineteenth century, something unconnected had happened which was to affect the future of *Terra Nova*. A young man had set himself up as a watchmaker, silversmith, engraver and jeweller. Benjamin Bowring began business in the south Devon city of Exeter. Later, he crossed the Atlantic in pursuit of greater ambitions ...

## 2

# ACROSS THE ATLANTIC AND NEW OWNERS

### The Ambitions and Achievements of Benjamin Bowring and his Family

With the end of the nineteenth century approaching, the 'march of time' was bringing change and *Terra Nova* was to have a different country, port and ownership for the rest of her life. Those early nineteenth-century ambitions of a Devon watchmaker turned merchant venturer would eventually affect her destiny.

The cultural and political development of Newfoundland as a colony was shaped by the cod fishery and later the sealing industry. For hundreds of years, Europeans had fished the Grand Banks and had visited Newfoundland and Labrador to fill their ships with dried and salted cod. With emigration to the North American continent, the population of Newfoundland grew, and activity increased in both industries. The Bowring company, controlled by a family dynasty, was becoming very much a part of this growing scene.

There was an increased demand from Britain for oil in the early nineteenth century and the export of seal oil began to match the trade in the export of salted fish. The large increase in Newfoundland's population became employed in the allied industries of seal skinning and the manufacture of casks and drums necessary for shipment. Companies grew larger and could afford to operate bigger ships.

Harp seals had always been plentiful around Newfoundland on a seasonal basis. The seal meat was consumed locally, while the pelt, with its thick layer of fat, was used to produce oil and leather. As the population of the colony grew and with larger, more powerful vessels available, the capability to move further north increased access to other breeds of seals.

When the sealing season was over, they turned to fishing. This increased activity extended to the Labrador coast where the fishing industry had been long established. The inadequacies of sail power were overtaken by the introduction of steam, enabling vessels to move more effectively in the ice fields. Two St John's firms, Walter Grieve & Co. and Baine Johnson & Co. were the first in Newfoundland to acquire the steamers *Wolf* and *Bloodhound* for the sealing industry. They were followed by Job Brothers, who acquired *Nimrod*. Ridley, Son & Co. acquired *Retriever* and John Munn & Co. acquired *Mastiff* and *Commodore*. All these vessels were built at Dundee by Alexander Stephen & Sons Ltd.

The company begun by Benjamin Bowring had now been established for nearly fifty years and was steadily growing. In 1865, Bowring Brothers bought the gunboat, HMS *Plover* (173 tons), and renamed her *Hawk*. She was Bowring's first steamship.

The future for *Terra Nova* in Newfoundland was becoming more certain. From 1898, the future working port for SS *Terra Nova* would be St John's.

## The Founding of a Shipping and Trading Company

When, in 1898, the SS *Terra Nova* passed into the ownership of Bowring Brothers at St John's, Newfoundland, she joined a company already well established under the title of C.T. Bowring and Company. They had offices in London and Liverpool and branches in several other cities. The origin of their maritime activities had begun in Newfoundland and from the time of their beginnings in 1816 the company had been energetically led by successive generations of family members. This saw the company grow with the Industrial Revolution, parallel with the development of the British Empire in the nineteenth century, through two world wars and well into the late twentieth century.

The 'New Found Land' on the east coast of the North American continent was discovered in 1497 by the explorer and merchant venturer, John Cabot. Less than a century later, in 1583, Sir Humphrey Gilbert, the Devon navigator, took possession of the colony and founded a settlement, thus making it the oldest British colony. It marked the beginnings of the British Empire. Newfoundland became a self-governing Colony of Great Britain and eventually joined the Confederation of Canada in 1949.

Water Street, St John's, 1837. (*The Bowring Company Magazine*)

From the fifteenth century, merchant venturers from the West Country had sailed vessels across the North Atlantic to the east coast of the 'New World', taking with them saleable goods in exchange for the rich and abundant dried salt cod produce of Newfoundland and Labrador. There followed the 1620 emigration of the Pilgrim Fathers and the subsequent growth of 'New England'. The rich cod-fishing grounds of the Grand Banks and the seal fishery of the North Atlantic coast off Newfoundland provided a natural resource.

The Latin words '*Terra Nova*' mean 'new land' and it seems appropriate that when the SS *Terra Nova* was so named in 1884 by her builders she should, for forty-five years from 1898 to her end in 1943, be destined to pass into the ownership of a shipping and trading company founded in the 'new land', Newfoundland. Early in the nineteenth century a young man, Benjamin Bowring, had set himself up as a watchmaker, silversmith, engraver and jeweller. His fourteenth-century forbears originated in the Kingsbridge area of the County of Devon where the family was engaged in the wool trade.

Benjamin Bowring was born at Exeter in 1778, where he became an apprentice watchmaker, eventually opening a business on his own account. Later, he crossed the Atlantic in pursuit of greater ambitions, for he had realised the potential of expanding his fortune in the growing cross-Atlantic trade.

He began his own business in a shop at Exeter situated nearly opposite St Martin's Lane in the High Street, adjacent to the ancient Guildhall. In 1803, he advertised his business in the *Exeter Flying Post*. In that year, he married the daughter of a watchmaker, Charlotte Price. Moving to nearby premises, he published further advertisements to promote his trade, seek an apprentice and to express his appreciation of customers in helping him to secure his successful business.

His ambitions led him to think of crossing the North Atlantic. Family legend has it that he may have been inspired to do so by a Newfoundland trader who visited Exeter and bought a clock at Benjamin's shop. There, the trader would have talked of his ship and cargo of cod. Other Bowring descendants had already emigrated to North America and kept contact with Benjamin as best they could. He disapproved of the slave trade and this had made him unpopular locally with those of opposing views.

The West Country trade with the 'New World' very much fired his ambitions and in 1811 he went to Newfoundland with the intention of expanding his business. St John's, at that time, was very much a fishing settlement, its main street just a series of wooden houses, stores and fish sheds. Between 1811 and 1815, Benjamin made a number of journeys to Newfoundland. On one occasion, the ship was captured by US privateers, who relieved him of his stock of clocks and watches and dumped him on the ice outside St John's, leaving him to make his own way to the harbour.

In 1815, an advertisement appeared in the *Exeter Flying Post* announcing the sale of his stock and thanking customers for their business and support. He had clearly decided where his future business lay.

In 1816, Benjamin Bowring, with his wife and three sons arrived in Newfoundland where their business was initially a continuation of the Exeter trade. An advertisement appeared in a Newfoundland journal on 14 September 1816, informing the residents of St John's that he was in business as a watchmaker, silversmith and jeweller as well as selling many other household goods. To support the business, his wife ran a general store.

St John's, at that time, consisted of closely built and hurriedly constructed timber properties. There were serious fires in 1816 and 1817, causing devastation to growing businesses, homes and families. These were hard times in Newfoundland, with increased immigration in the period following the wars with the American states and France.

However, the family soon adapted and in addition to watch and clock making they turned to the trade of the colony, supplying fishermen with their requirements and exporting their produce. Benjamin acquired his own wharf and purchased two schooners, *Charlotte* (44 tons) and *Eagle* (91 tons). He also

Water Street, St John's, around 1890. (Courtesy Bridie Malloy's Bar, St John's)

bought 3,000 acres of land. He traded with Bristol and Liverpool and with relatives in Devon and other parts of the country who were in the wool trade as serge makers.

In 1819, a fourth son, Edward, was born, then John in 1824, who was followed by a daughter, Charlotte. But in 1828, there was a family tragedy, for their first son William, aged 24 years, was lost at sea. In 1833, the premises of Benjamin Bowring & Son were destroyed by yet another fire. They were only partly insured, and Benjamin wrote to his agents in Liverpool assuring them that he would honour his debts despite the setbacks.

The severe losses almost led to the whole family leaving Newfoundland but, confident there was a good business that could prosper, they were inspired and felt committed by the return from England of their now eldest son, Charles Tricks Bowring, who had brought his new wife, Harriet, with him to St John's to work in the business.

In 1834 Benjamin Bowring, now 56 years of age, with his wife, youngest son and daughter, left Newfoundland for Liverpool. This left three sons in Newfoundland to run the business, which traded as Bowring & Son until 1839 when it became known as Bowring Brothers. In 1840, they purchased the brig, *Margaret Jane* (103 tons) to carry the firm's first load of sealskins and other cargo to Liverpool.

The parent company title of C.T. Bowring & Company, originates from 1841 when, following his arrival in Liverpool, Benjamin Bowring established a business at King Street, Liverpool. He bought a brig, *Velocity* (145 tons) and in turn began trading with Newfoundland.

When he died at Liverpool in 1846, aged 68 years, Benjamin Bowring, the merchant venturer, was a respected citizen at the port, who had contributed much to the early commercial development of Liverpool and to St John's, Newfoundland. The business was now in the hands of his eldest surviving son, Charles Tricks Bowring, after whom the company was then named.

Within weeks of the death of Benjamin Bowring, another devastating fire struck St John's and the buildings of Bowring Brothers Ltd were again destroyed. Once again, by sheer determination, the difficulties were overcome and in 1854 orders were placed to acquire more ships for the international trade. By 1858, nine ships were owned or chartered and in 1861, seven ships were sent to the sealing trade. In 1870, the Bowring companies could list twenty-two ships at St John's and fifteen at Liverpool.

Charles Tricks Bowring JP (1808–85). (Courtesy the Bowring family)

They bought their first steam ship in 1865 and with the advance of steam power in ships, in 1884 more steam vessels were added to the fleet to operate a passenger and freight trade down the eastern seaboard of North America to

North side of St John's Harbour and Bowring Brothers premises. (Courtesy of Bridie Malloy's Bar, St John's)

Halifax, Nova Scotia and New York. The company line insignia on the ships' funnels and on the house flag was a long, red diagonal cross on a white background. The company was known as the Red Cross Line.

In 1892, yet another devastating fire struck St John's and the whole town was virtually destroyed. It was estimated that 10,000 people were made homeless; international relief funds were set up. The company, however, survived. Gradually, over the years each of the brothers returned to live in Liverpool, leaving the company to be managed by successive generations of the Bowring family.

Charles Tricks Bowring died at Liverpool in 1885 and was succeeded as head of the firm by his brother, John, whose chairmanship was short-lived for he died the following year at his mansion home in Liverpool, which he had aptly named, 'Terra Nova'.

From those early beginnings in 1816 to the eventual sale of their fleet later in the twentieth century, the company had owned 200 ships. The company had known many shipping disasters and during both world wars lost ships to enemy action.

When, in 1898, the SS *Terra Nova* joined the fleet of Bowring Brothers at St John's, the business of Bowring Brothers was managed by Charles, grandson of the founder and the second son of Charles Tricks Bowring. By the late nineteenth century, the parent company's interests were growing worldwide with entry into the advancing oil and petroleum industry with orders placed for larger ships and tankers to carry bulk oil and fuel rather than in barrels

The town of St John's, north side of the harbour after the devastation of the great fire of 1892. (Courtesy Rosalind Power)

SS *Terra Nova*, leader of the wooden fleet, frozen in at St John's Harbour around 1900. (Courtesy Memorial University, Newfoundland)

on sailing ships, as in the early days. In 1903, the company entered into the marine insurance and reinsurance market.

In *Lloyds Register*, 1912–13, the telegraphic addresses for C.T. Bowring & Co. were given as:

20 Castle Street, Liverpool.
Winchester House, Old Broad Street, London.
Water Street, St John's, Newfoundland.
Atlantic Buildings, 59 Mountstuart Square, Cardiff.
Whitehall Buildings, 17 Battery Place, New York.

In 1980, 164 years after being founded by Benjamin Bowring, C.T. Bowring & Co. Ltd and their associate companies became the target of an aggressive takeover bid by the New York-based insurance company of Marsh McLennan, determined to strengthen their position in the world insurance market. Despite much opposition and efforts to save this very old British company, Bowring's shareholders found the offer too attractive and the company changed hands for $580 million. All Bowring's ships were eventually sold, the new company's main interest being the international insurance and reinsurance market.

These events were a long way from the thriving and expanding days of the traditional British family-led company which acquired the SS *Terra Nova* toward the end of the nineteenth century, adding her to its growing fleet of merchant and passenger ships.

However, let us go back to the days of 1898 and her acquisition by Bowring Brothers. From then to 1902, *Terra Nova* was under the command of Captain Arthur Jackman, a member of a well-known family of mariners, a skilled navigator and sealing master. She was operating with a fleet of similar vessels, all making an important contribution to the economy of the province in terms of seal products and employment.

Although 1898 is given as the year she was sold to Bowring Brothers Ltd, clearly for many previous years there had been much co-operation between the port of Dundee and Newfoundland with the chartering and manning of ships which kept in production the processing plants leased by the Stephen company at St John's. As far back as 1889, Captain Charles Dawe, Newfoundland men are recorded as commanding *Terra Nova*, indicating that prior to her sale she was chartered by her Dundee owners to Newfoundland operators. Further, the autobiography of Captain Abram Kean shows that he was captain of the Bowring-owned *Aurora* for eight years from 1898.

Because she was well known in Britain and across the Atlantic as a large and powerful ship suitable for ice navigation, the SS *Terra Nova* was purchased in 1903 from Bowring Brothers by the Admiralty to assist in the relief of the British National Antarctic Expedition of 1901, led by Commander Robert Falcon Scott RN.

## 'Terra Australis Incognita'

For her first adventure in polar exploration, *Terra Nova* was to go south. Governments, scientific institutions and private sponsors in many countries had become interested in the Antarctic and were supporting expeditions. None wanted to be left behind in exploring and making discoveries on this new frontier, but it was the whalers and sealers who had initially pushed the boundaries both to the north and south. *Terra Nova* had now been acquired as a relief ship to assist a British expedition already in Antarctica.

Enthusiasm for polar exploration by the British had been dampened by the events of the Arctic Expedition of Admiral Sir John Franklin and the search

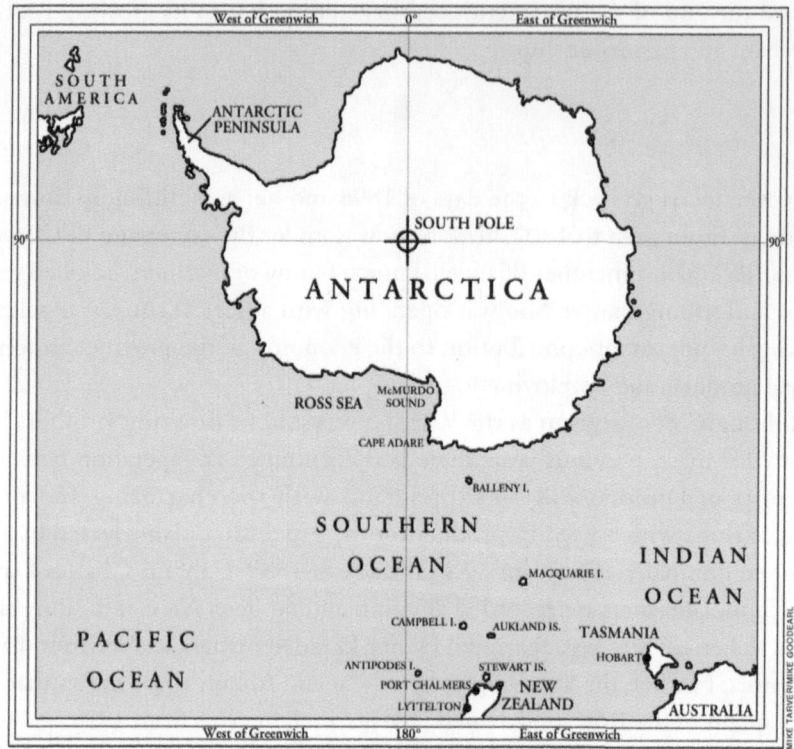

for the Northwest Passage in 1845, when the whole expedition had been lost. However, the President of the Royal Geographical Society, Sir Clements Markham, a former Royal Naval officer himself, had long nursed a desire to explore the Antarctic and he had expressed his indifference to the British Expedition led by Carsten Borchgrevink. He dismissed it as being conducted by foreigners and not properly British.

Less than 100 years ago and into the twentieth century when there were no radios, no helicopters and no synthetic materials, the north and south extremities of the Earth were yet to be reached and explored. Now SS *Terra Nova* was to play a part in these exciting adventures as a relief ship to assist the RRS *Discovery*. The British Antarctic Expedition led by Commander Scott RN had left Britain in 1901 and reports brought back by the relief ship suggested that *Discovery* was now trapped by ice and might be unable to release herself.

These events were part of the many sagas which arose during the exploration of both polar regions. The honour of being first to cross the Arctic Ocean and reach the North Pole continues to be in dispute. Was it either of the US' explorers Robert Peary or Frederick Cooke, both or neither? Each claimed success.

A greater mystery was the huge continent in the south. Antarctica is larger than Europe or Australia. Though something was known to be there, the very early astronomers and other experts of their day speculated that there had to be as much land in the southern hemisphere as in the north, 'otherwise the spinning world might be unbalanced and thrown off its axis'.

Sir Francis Drake in his famous sixteenth-century round-the-world voyage had sighted ice in the south. There had been some important land falls. The Dutch explorer, Abel Tasman discovered new territory in 1642 and named it Van Diemen's Land after the governor of the Dutch East Indies. It was renamed Tasmania in 1853. The English navigator and venturer, William Dampier landed on Australia in 1685 and charted part of the north western coast. However, no one knew exactly what was there. Was it one continent or an archipelago? It was charted as '*Terra Australis Incognita*'.

There were problems for the navigators, too. As ships went further south compasses became inaccurate. Ships were going hundreds of miles off course because of the differences now known between the geographic and magnetic poles, caused by the Earth's magnetism. Latitude was understood, but longitude was still a problem to determine and was to be the great scientific puzzle solved by the clockmaker, John Harrison. He eventually won the prize of £20,000 grudgingly awarded by the British Government Board of Longitude in 1773 for his making of the timepieces H1–4 which can be seen today at the Royal Observatory, Greenwich.

Captain James Cook RN, of Whitby, Yorkshire is really the man who began the discovery of the southern hemisphere. He had charted Newfoundland, the Labrador coast and the St Lawrence Estuary enabling an amphibious assault by British forces under Brigadier Wolfe to defeat the French at Quebec in 1759. In 1768, with his talents acknowledged, Captain Cook was sent south by the Admiralty in HMS *Endeavour* and charted New Zealand, sailing through the straits which now bear his name. He had aboard with him a Harrison clock and other timepieces of the day. In his second voyage south in 1772, aboard HMS *Resolution* and accompanied by HMS *Adventure*, James Cook continued his work and charted much of the remaining coast of Australia, only to find his way further south to be blocked by ice.

As the early whalers and sealers from different countries pushed their way further south, it was Captain James Cook in the *Resolution* and *Adventure* which were the first ships to cross the Antarctic Circle on 17 January 1773. On the following day, heavy ice conditions caused them to turn north again. Much of the mystery of '*Terra Australis Incognita*' had been solved. The important information was not what Captain James Cook had discovered, but what theories he had disproved. He had shown that the south consisted of continents and islands that he had charted and further south, there lay a frozen secret.

Tragically, in Hawaii on his third voyage he was killed on a beach by islanders after a quarrel over a longboat that had been stolen from his ship. Today, he is regarded as the father of marine navigation. He did much to improve the standards of living at sea, including the control of scurvy and other deprivations aboard ship.

With Australia and New Zealand added to the world map the mystery of the frozen south became a challenge. The first to sail these waters were the whalers and the sealers who brought back stories of ample stocks for the taking. As these stories spread, ships of many nations began to move south.

In 1841, Captain James Clark Ross was sent south by the Admiralty in HMS *Erebus*, accompanied by HMS *Terror*. Their mission was to find a way through the pack ice and discover what was beyond. This, in sailing ships, they amazingly did and entered the sea which is now named the Ross Sea. Captain Ross and his men were the first to see the two volcanoes, which he named after the two ships of the expedition. He named McMurdo Sound after the 1st officer of HMS *Terror* for his skill and zeal in seamanship.

However, penetration inland of the Antarctic continent was yet to begin and that brings us to the turn of the nineteenth/twentieth century. History reveals claims for a number of landings on the continent and among them was a party from a whaling and sealing ship, the *Antarctic*, an expedition led by Henrik Johan Bull, who raised the Norwegian flag at Cape Adare on 24 January 1895.

The first wintering below the Antarctic Circle was made by an expedition led by a Belgian, Baron Lieutenant Adrien Victor Joseph De Gerlache de Gomery in the *Belgica*. This took place in 1898 when the ship became accidentally trapped by pack ice in the area west of that now known as the Antarctic Peninsula. Some of the crew went mad. Aboard that ship was the Norwegian, Roald Amundsen.

The first man to raise the British flag on the continent was Carsten Borchgrevink, leading the British Antarctic Expedition of 1898–1900, in the ship, *Southern Cross*. Borchgrevink was a Norwegian living in Australia who had raised money in London, mainly from the publisher Sir George Newnes. His expedition was manned by British, Norwegians and Laplanders. Borchgrevink had been there before, he was one of the party from the *Antarctic*, who stepped ashore at Cape Adare to raise the Norwegian flag in 1895.

Thus, SS *Terra Nova* found herself drawn into the mystery of polar exploration and discovery. Her first involvement was the Antarctic continent.

# 3

# FROM THE NORTH ATLANTIC TO THE ANTARCTIC

Relief Ship for the British National Antarctic Expedition

In 1903, *Terra Nova* was purchased by the Admiralty from Bowring Brothers Ltd to act as a second relief ship for the British National Antarctic Expedition, 1901–04, which was led by Commander Robert Falcon Scott RN, who was currently iced in on board the RRS *Discovery*.

Following the first relief of the *Discovery* expedition by the *Morning* in January 1903, the organisers received first-hand reports from Captain William Colbeck of *Discovery*'s position iced in at McMurdo Sound. After much debate by the promoters of the expedition, the Royal Society and the Royal Geographical Society, the decision as to whether or not *Discovery* should remain at her base during the Antarctic winter was to be left to the judgement of the expedition leader and captain of the ship.

In this respect, Scott was fortunate. Not only was he leader of the expedition, but he was captain of his ship. Not so with many polar expeditions to the polar regions, north and south, which suffered from conflict between the risky ambitions of expedition leaders and the heedfulness and caution of ships' captains.

Now, with the news brought by Captain Colbeck and others on *Morning*, who had returned from the Antarctic, the debate was reopened. First, there was the financing of a third winter for the expedition – should *Discovery* remain iced in? Further, for just how long might she be iced in and what might be her fate? Discussion of the possibilities brought fear of another disaster along the lines of the lost Arctic Expedition of Sir John Franklin, fifty years earlier.

Captain and crew of *Terra Nova* for the relief of RRS *Discovery* 1903–04. Captain Harry McKay is in the back row, fourth from left. (Courtesy Scott Polar Research Institute)

The two societies had limited resources and, with Royal Navy personnel involved, the situation demanded definitive action. The Admiralty, having been consulted, took full control of the situation, overriding the two scientific societies. The Admiralty bought *Morning* from the societies and, in addition, acquired a larger and more powerful ship. They purchased *Terra Nova* from Bowring Brothers, sending both ships with orders to Scott that if *Discovery* could not be released from the ice she was to be abandoned and the expedition brought home.

## Under Dundee Command Again and First Mission to Antarctica

*Terra Nova* was put under the command of a Dundee man, Captain Harry McKay, and fitted out at the Panmure yard in record time. She was crewed almost wholly by Dundee men (for the crew list, see Appendix E). The occasion was recalled nearly fifty years later by crew member, Thomas Cosgrove, aged 75 years, of 52 Kemback Street, Dundee, in an article published in the

*Dundee Evening Telegraph*, dated 31 March 1951. He had served in *Terra Nova* during her whaling days and, knowing the ship well, he signed on for the relief voyage.

The ship departed for the south on 21 August 1903. She was towed by a succession of Royal Navy ships through the Suez Canal and the Red Sea to the Indian Ocean. *Terra Nova* met *Morning* off Hobart, Tasmania. Both ships sailed together for the Antarctic, reaching McMurdo Sound on 5 January 1904.

Scott had long been trying to deal with the predicament of having his ship still trapped by the ice barrier which extended up to 20 miles north of Hut Point. Attempts had been made to saw cuts in the ice in an effort to obtain a release, but this fruitless exercise was

Thomas Cosgrove of Dundee. (*Evening Telegraph*, 1951)

abandoned. On 5 January 1904, Scott and Wilson, watching wildlife, were camping on the ice barrier near Cape Royds, 3 miles from the edge, when the two relief ships came into view.

Much to Scott's surprise, he saw two relief ships, although did not realise the full implications. He later recorded:

> Looking back now, I can see that everything happened in a such a natural sequence that I might well have guessed that something of the sort would come about, yet it is quite certain that no such thought ever entered my head and the first sight of the two vessels conveyed nothing but blank astonishment.

They had expected *Morning* but did not know the reason for the other ship's presence or her identity. Arrangements were made for a dog team to collect mail from the ships. Scott and Wilson made their way to the ice edge to meet the new arrivals, to learn that the other ship was *Terra Nova* and her captain, Harry McKay.

Once again, Scott met the captain of *Morning*, William Colbeck, who had relieved the expedition in 1903. It was then that the full extent and implications of this final relief fell upon Scott. What really concerned him at the time was the content of the Admiralty instructions from London, handed to him by

Freeing of *Discovery*: *Terra Nova* is in the centre and *Morning* is to the right. (Courtesy Scott Polar Research Institute)

Captain Colbeck. This placed Scott and his expedition in a cruel position. It was all a question of pride, because if the 20 miles of fast ice between *Discovery* and the open sea did not break up within six weeks they would be forced to abandon their ship and return home – 'as castaways, with the sense of failure dominating the result of our labours', wrote Scott.

Here, it can be said that the rapid and vicious changeable weather conditions on the continent helped play their part. Explosive charges were used to break open the ice, assisting both rescue ships to get nearer to Hut Point. Thomas Cosgrove, aboard *Terra Nova*, recalled the sight when the ship neared the ice barrier:

> It varied in thickness from 13 to 15 feet thick. We climbed on to the ice and dug foot-square holes. In these we put explosive charges. It took us the whole day, then at night we blasted. These operations were combined with ramming the ice with the ship's bow, going astern and then going full speed ahead.

When *Terra Nova* reached *Discovery*, Tom recalled that he and another man were the first to clamber aboard to be greeted by Captain Scott, who said, 'Good old Dundee!' He shook hands with them and a huge silk Union Jack was hoisted.

From the time of their arrival on 5 January to 14 February 1904, the relief ships worked their way slowly and steadily towards *Discovery*. Occasionally there were changes in weather conditions which assisted them but the real breakthrough came on 14 February, when a strong wind from the south-east caused the ice to break away north, allowing both relief ships to steam up to Hut Point.

The following account of the final breakthrough by *Terra Nova* was given by an unnamed officer aboard the ship and appeared in the newspaper for Central Scotland, *The Courier*, on 23 April 1919:

> About 4.30 p.m. we began 'butting' and did an hour at it. At six p.m. we resumed and Captain McKay remarked that his chance had now come and he would take it. There was a lead of open water off Hut Point and it was our object to burst through the intervening ice and get up there. Once there we should be quite close to the still ice-bound *Discovery*. As we kept on 'butting' the *Discovery*'s crowd began to gather round the flagstaff at Hut Point and watched us eagerly as we gradually but persistently got near to the goal. The *Morning* was on our port side and was kept busy poking into the cracks which we made when we came full speed up against the solid floe. we went on 'butting' all hands, as before, 'rolling ship.
>
> The *Terra Nova* did some really good work and on one specially well remembered occasion she broke a big piece off on her port side. We kept on ahead chewing the ice up, then almost stuck for a little, when away we went crashing through the last few yards of ice and entered the open water at Hut Point about 10.30 p.m. on February, 14th. As soon as we got through the crowd on Hut Point cheered lustily and hoisted the Union Jack, while all aboard our ship gave three ringing cheers for Captain Harry McKay for the good old ship and for the *Morning*. The *Morning* soon followed and both ships tied up to the fast ice not many hundred yards from the *Discovery*. It was thought that the *Terra Nova* would never break through the ice to reach the *Discovery*, but Captain McKay was determined from start to finish. As one of his men said at the time, 'If the Discovery were in hell, he'd get to her'.

Explosive charges were placed fore and aft of *Discovery* and finally, on 16 February, she was free and ready to put to sea again. With the *Terra Nova* warped alongside *Discovery*, Captain Harry McKay and his officers were invited aboard that night for dinner and the merry party sat down to enjoy an Antarctic feast to celebrate.

No sooner had they sat down, than word came that a wind had sprung up. Scott assessed the situation and reluctantly had to inform his guests that they

Looking aft over the decks of *Terra Nova* (right) and *Discovery* warped together, starboard side, alongside Hut Point. (Scott Polar Research Institute)

were in for a stiff blow. Harry McKay and his men leapt over the rail to their ship without delay. With steam up, they cast off and were soon away to ride out the conditions in the open sea.

Arrangements had already been discussed for the three ships to depart McMurdo Sound, but priority was for the coal stocks needed for each ship on the return to New Zealand. *Terra Nova* transferred 50 tons to *Discovery*, and she received a further 25 tons from *Morning*.

It had been decided that the three ships should sail together as they went north up the coast, then the smaller and less powerful *Morning* should be free to begin her passage home directly north, the other ships going north-west. *Discovery* had a precarious start to the departure from her berth at Hut Point, being blown on to a lee shore with a receding sea brought about by the conditions. In a day that almost drove them to distraction, having released themselves from the ice they found the ship being grounded on to the beach immediately below Hut Point, her keel striking the shoal beach with sickening thuds.

Eventually, the weather changed and to their relief the ship came off astern. (Another illustration as to how weather conditions can change rapidly and alter any Antarctic scene, creating accompanying dangers.) On this occasion, the forces of nature came favourably to their rescue and no serious damage

The relief ship, SY *Morning*. (Courtesy Dr Hugh Russell MD, Edinburgh)

was caused to the ship as far as they could tell at that time.

*Terra Nova* went alongside *Discovery* again and they warped together off Glacier Tongue, north of Hut Point where they transferred more coal. Here *Morning* rejoined them and the transfer of coal to *Discovery* was completed.

The passage to New Zealand was discussed and it was agreed that all three ships would rendezvous at Port Ross in the Auckland Islands and sail together for New Zealand, all arrangements being made for the usual possibilities and eventualities.

Cutlery from relief ship, SY *Morning*. (Michael Tarver, courtesy Hon. Edward Broke Evans)

Crew of relief ship SY *Morning* (left to right): 3rd Officer Gerald Doorly, 2nd Officer Teddy Evans, Chief Engineer Frederick Morrison, 1st Officer Rupert England, Captain William Colbeck, 4th Officer Sub lieutenant George Mulock, Surgeon George Davison (seated above), Midshipmen Maitland Somerville and Arthur Pepper. (Courtesy the Hon. Edward Broke Evans)

On 21 February 1904, *Morning* sped on her way north, taking advantage of a good wind, *Discovery* and *Terra Nova* made their way north up the coast towards Cape Adare. On the 24th, *Terra Nova* received a signal from *Discovery* that her rudder was damaged, possibly caused during the grounding ashore or on the passage north. Both ships made for Robertson Bay where the damaged rudder was lifted and the spare fitted. While work was being carried out, the tidal flow sent heavy ice floes toward them. *Terra Nova* left the bay hurriedly. *Discovery* had no option but to remain at anchor while work continued.

The following day, work fitting the spare rudder was complete, *Discovery* rejoined *Terra Nova* and soon both were steaming north-west. On 28 February, the ships lost contact with each other until their arrival at Ross Harbour in the Auckland Islands. *Discovery* arrived on the 15 March ahead of *Terra Nova*, which arrived on 19 March, and the following day *Morning* arrived safely after a very difficult passage. Tales were exchanged telling of the difficulties each ship had encountered on their way north and a well-earned rest of ten days was enjoyed.

On 29 March all three ships sailed together towards Stewart Island and the coast of the mainland, favoured with a good wind and a moderate sea. They arrived at Lyttelton on 1 April 1904, and once again were met with a warm welcome and much kindness from the people of New Zealand. Scott, his expedition members and the *Discovery* became celebrities and news of their return and achievements was made known around the world.

*Terra Nova* was put into dry dock before her return to the UK. She set her course home via the Falkland Islands and arrived at Sheerness on 18 August 1904, a month ahead of *Discovery*.

*Discovery* arrived back at Spithead on 10 September and, later, on 15 September she berthed at East India Dock, London. *Morning* arrived back in the UK at the end of September.

On the return of *Discovery* to Britain, the British National Antarctic Expedition was acclaimed an outstanding success, both scientifically and territorially. Much new material was brought back for evaluation and important sledge journeys inland had been made, including a journey 'farthest south' – further than man had ever been before. Its leader, Commander Scott was acclaimed a national hero and promoted to the rank of Captain RN.

However, while the nation revelled in the exciting news and admired their new heroes, controversy continued over the use and cost of the relief and *Terra Nova* lay out of the 'limelight' at her berth in the West India Dock. It was not to be for long, however, as she was destined for another polar mission, but this time in the north and the Arctic Ocean.

# 4

# FROM THE ANTARCTIC TO THE ARCTIC

## Under American Ownership and Norwegian Command

*The Ziegler Arctic Expeditions*
The Bowring Company records of activities do not show *Terra Nova* undertaking sealing during 1904. Due to her involvement in the south, she was too late for the sealing season and we can assume that she remained in the UK during the months between August 1904 and January 1905, when she was purchased by William Ziegler, an American businessman and millionaire.

William Ziegler promoted two Arctic expeditions in an attempt to reach the North Pole. These expeditions were among numerous attempts by explorers from different countries to penetrate the Arctic regions by means of Smith Sound, Greenland, Ellesmere Island, Svalbard and Franz Joseph Land. All had hopes of being the first to cross the Arctic Ocean to reach the North Pole.*

*(1) The Baldwin–Ziegler Arctic Expedition, 1901–02*
William Ziegler, who had made his money in baking powder, was seriously ill and he wished to promote a worthwhile project for which he could be remembered. He had become consumed by the ambition for an expedition to reach the North Pole. He wanted to be famous for more than baking powder and financed the Baldwin–Ziegler Arctic Expedition of 1901–02. He

---

\* Some publications state that *Terra Nova* relieved the Jackson-Harmsworth Arctic Expedition 1894–97. This is inaccurate. She had no connection with that expedition.

is reputed to have invested more in the cause of Arctic exploration than any other man in the world.

This expedition consisted of three ships, *America*, *Frithjof* and *Belgica*. *America* was the former British whaler, *Esquimaux*, purchased and renamed for this expedition. (*Esquimaux*, at 593 tons, was another ship built by Alexander Stephen & Sons and is No. 33 on the list, launched at the Dundee yard for the Dundee Seal & Whale Fishing Co. in 1865).

Ziegler chose as leader of this hugely expensive expedition Captain Evelyn Briggs Baldwin, a US meteorologist. Baldwin had earlier been involved in Robert Peary's Greenland Expedition, 1893–94, and in another attempt to reach the North Pole from Franz Joseph Land by Walter Wellman in 1899.

Baldwin sailed in *America*, which was commanded by a Norwegian, Captain Carl Johansson, and was supported by the *Frithjof* under the command of Captain Johan Kjeldsen. The other ship, *Belgica*, was to lay depots in Greenland for the proposed return route.

William Ziegler, founder of the Royal Baking Powder Co., USA. (Courtesy Martha's Vineyard Historical Society)

On arrival in the region, Captain Kjeldsen, after great difficulty, finally located the leading ship, which resulted in a furious row between Kjeldsen and Baldwin. Baldwin, the leader, was already in conflict with the captain of his own ship.

These incidents were the result of bad planning, misplaced equipment, lack of co-ordination and poor leadership. This led to the termination of the whole expedition, which returned a complete failure. The result was bitter recriminations and threats of legal proceedings between Ziegler and Baldwin.

*(2) The Fiala–Ziegler Arctic Expedition, 1903–05*

The expedition sponsor, William Ziegler, was now terminally ill but still determined to achieve his goal. He financed a second expedition in an attempt to reach the North Pole, again from Franz Joseph Land. Just one ship from the previous expedition was reused, the *America*.

Captain Edwin Coffin, Fiala–Ziegler Arctic Expedition, 1903–05. (Courtesy Martha's Vineyard Historical Society)

This time, Ziegler appointed Anthony Fiala as leader. Fiala had been a member of the 1901–02 expedition and after it failed Ziegler questioned some of its members and met Fiala, the 32-year-old expedition photographer.

Fiala was a military man, artist, adventurer and soldier of fortune as well as photographer and Ziegler was impressed with him. Ziegler put William S. Champ, secretary and former sales executive in the Royal Baking Powder Company, in charge of managing expedition affairs.

Scientific exploration was to be secondary to the honour of gaining the main objective, which was to be the first to arrive at the North Pole and there place the Stars and Stripes of the USA. Captain Edwin B. Coffin, aged 52 years, of Edgartown, Massachusetts, was appointed captain of the expedition ship. He was a career sea captain with much experience of ice floes and the Arctic Ocean.

The expedition sailed from Vardo, Norway on 10 July 1903. America had a tortuous voyage through the ice, reaching southern Franz Joseph Land on 12 August 1903. The expedition consisted of thirty-eight men, 218 dogs and thirty ponies. They pressed on through the archipelago, reaching Teplitz Bay on the northernmost tip. This was the previous base of an Italian expedition led by the Duke of Abruzzi, who made an attempt on the North Pole during 1899–1900.

They pressed on north for a further 26 miles north of Rudolph Island before coming on heavy ice fields at 5 a.m. on 31 August. By going this far north, Coffin claimed that he had broken the record for a ship travelling north, i.e. to Lat. 82°14' N, just 525 miles from the North Pole. As it transpired, that was the furthest north the expedition was ever going to reach. They then turned back south to Teplitz Bay to make a base ashore.

Once again, there were the inevitable disagreements between the expedition leader, Fiala, and Coffin, *America*'s captain, as to the position to be chosen for a safe anchorage. Relations between the two men had not been good during their voyage north but the most serious dispute came when they arrived at Teplitz Bay.

Fiala and Coffin wrote memoranda to each other, Coffin pointing out the dangers of an exposed anchorage and the danger of ice floes coming in with south-southwest to north-north-west winds. He recommended an anchorage 8 miles south off Coburg Island and wished to accept no responsibility for the ship and its cargo if they remained in Teplitz Bay.

Fiala insisted that the ship remain in Teplitz Bay. He stated that he wished to commence his attempt north from this point and that to move the ship away would mean dividing the expedition and all its equipment, creating obvious difficulties. He gave orders to Captain Coffin to remain at Teplitz Bay. It was a decision that Fiala would regret.

Anthony Fiala (leader), Fiala–Ziegler Arctic Expedition, 1903–05. (Courtesy Martha's Vineyard Historical Society)

On 22 October, *America* was blown from her anchorage by gale-force winds. It took two days to recover the situation and return to the base. Ice pressure then began to build up round the ship and on 12 November came the first signs of serious trouble. The ship listed to starboard and was being crushed against the solid ice.

Fiala was now living aboard and recorded:

> On the morning of the 12th November, I was awakened about four o'clock by the shaking and trembling of the ship. I lay for some minutes listening to the groaning and moaning of the timbers under pressure of the ice and then 'Moses' the Captain's dog, pushed his way into my cabin and put his paws on me, looking into my face with his great black eyes as if beseeching me to rise. I learned later that after coming into my room he went below into the Captain's cabin and awoke him.

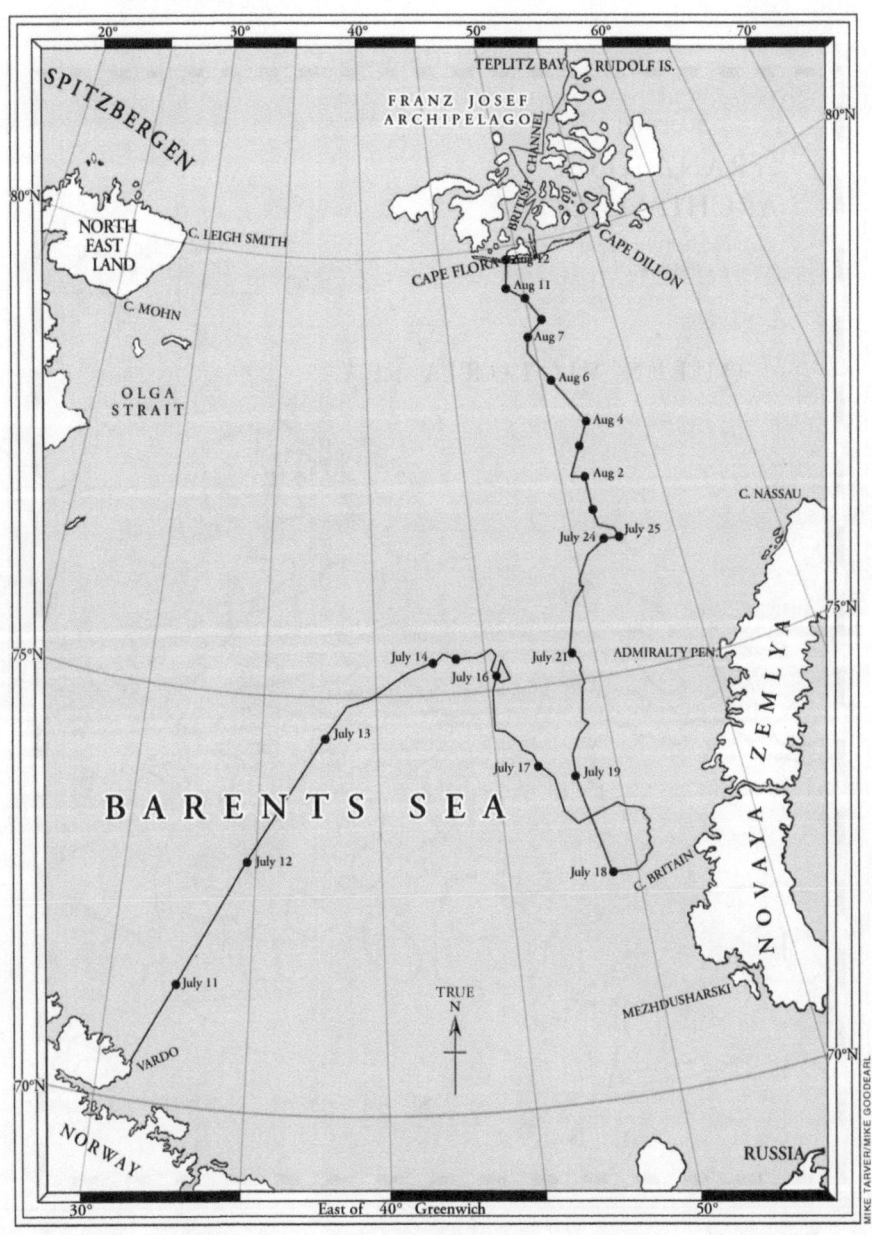

Fiala–Ziegler Arctic Expedition, 1903–05. Passage of *America* from Vardø, Norway, to Teplitz Bay, Franz Joseph Land. Reproduced from the personal papers of Captain Edwin Coffin. (Courtesy Martha's Vineyard Historical Society)

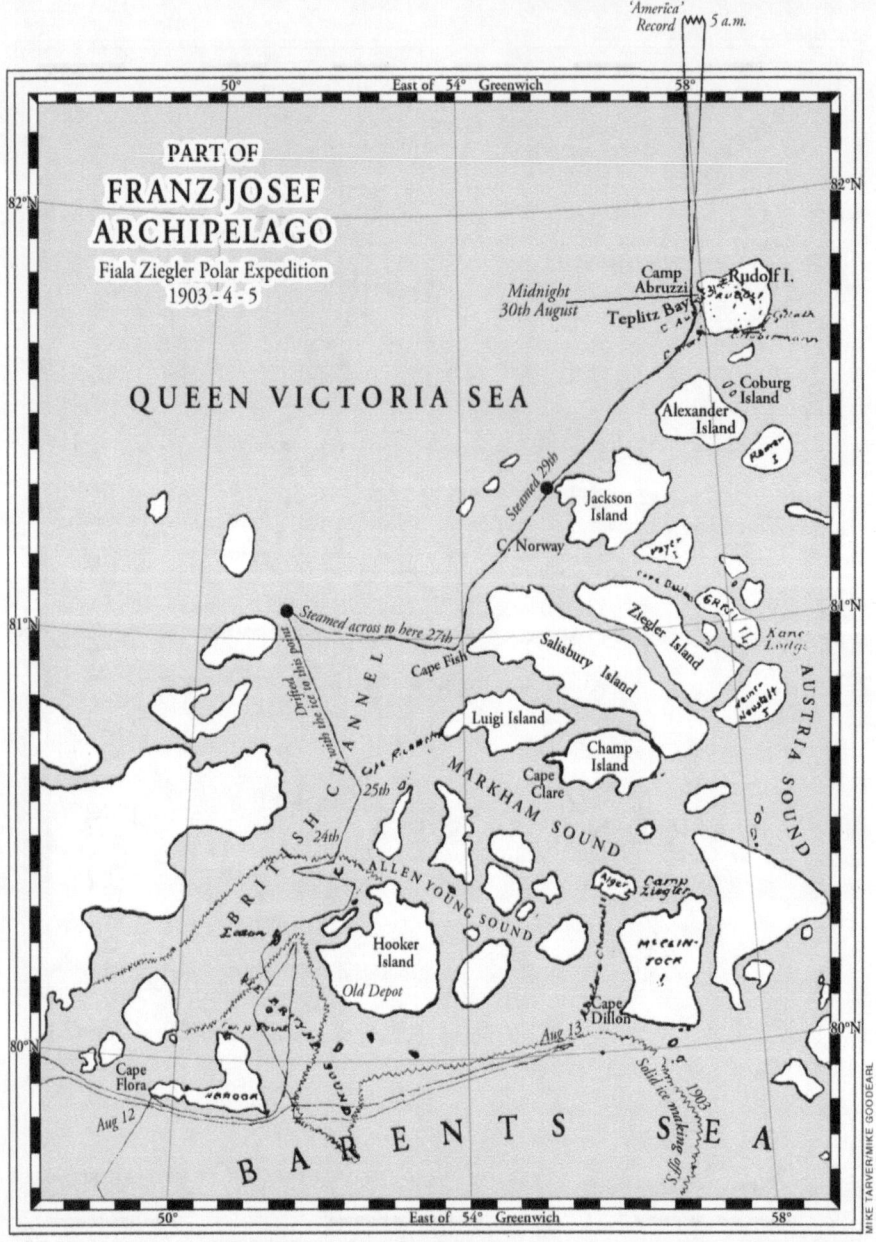

Fiala–Ziegler Arctic Expedition, 1903–05. Passage of *America* through Franz Joseph Land from Cape Flora to position north of Teplitz Bay where Captain Coffin claimed Lat. 82° 14' N as a record. Reproduced from the personal papers of Captain Edwin Coffin. (Courtesy Martha's Vineyard Historical Society)

On 21 November, *America* was finally crushed and sank at her disputed anchorage. The whole expedition was now accommodated in their camp hut. Materials from the wreck were used to enlarge and improve the meagre facilities. Three unsuccessful attempts were made on the North Pole, but all came to nothing, each failing because of the condition of the ice surface and the impossible task of being unable to cross the wide leads between the ice floes with the limited equipment available to them.

Now without a ship, the expedition members had to reorganise to travel the 165 miles to await eventual relief at the southern end of the archipelago. The expedition was then divided and a party prepared to leave for the south of the archipelago to the bases at Camp Ziegler and Cape Flora. The party that was to remain in the north made another attempt toward the North Pole, but that too was a failure and they returned to base camp after two days. Eventually, all members of the expedition made their way to the south of the archipelago to await a relief vessel.

Throughout the summer of 1904 they expected to be relieved and there were numerous exciting 'sightings' of ships. As August was ending, they began to realise that relief was not coming and they would have to prepare for winter. Added to this was the sad news that one of the men had died. Segurd Myhre, a Norwegian by birth and fireman on the *America*, had died of some unknown disease. He was buried at Camp Abruzzi on a knoll near the observatory. He was placed in a pine coffin and the grave marked with a cross. No medical help had been left for the party which had remained in the north and nothing could be done for him. He left a wife in Bergen.

The expected relief ship, *Frithjof*, a Norwegian whaler of 270 tons under the command of Captain Johan Kjeldsen, had attempted to reach them during the summer of 1904 but bad ice conditions prevented her progress, tearing iron plates from her bow. The expedition would have to wait for another year for a relief attempt.

*Terra Nova* was berthed in the UK awaiting orders, having returned from the Antarctic on 18 August 1904, after the relief of RRS *Discovery*. Her reputation as a relief ship was now well known and it was fortunate that she should at this time be back in the northern hemisphere in the UK and available for purchase by William Ziegler for another relief voyage.

Another of Alexander Stephen & Sons ships, *Thetis* (723 tons and built in 1881), had been acquired by the US Government to relieve the American Greely Arctic Expedition of 1881–84. Now *Terra Nova* was called to relieve another US Arctic expedition. Clearly, British polar exploration ships were better than those the US had available.

Ziegler, his health deteriorating, was concerned that the attempt by *Frithjof* to relieve the expedition during the summer of 1904 had failed, and it was fortunate that the more powerful *Terra Nova* was available in the UK awaiting orders to return to her home port. Now, as it turned out, she was to go to the Arctic, even further north than her normal sealing waters. She was bought from Bowring Brothers Ltd by Ziegler in January 1905, for $48,000. Now under US ownership, she was to be commanded by a Norwegian.

## Rescue Mission to an Arctic Archipelago

*Terra Nova* had sailed to Antarctica from northern waters and now she was to play her part in relieving an expedition which was seeking to extend man's conquest and knowledge of the Arctic region.

As far back as the fifteenth century, sailors started to search for a Northwest Passage, and the earliest exploration of land in the Arctic region is credited to Sir Martin Frobisher, Henry Hudson, John Davis, William Baffin and Sir Alexander Mackenzie. With others, their names are recorded on the charts.

In the nineteenth century there were further advances by Sir John Ross and Sir William Parry. In 1845, HMS *Erebus* and HMS *Terror*, previously used in the Antarctic by Captain James Clark Ross, were both refitted, and each had an auxiliary steam engine installed. Using these two ships, Admiral Sir John Franklin, with an expedition of 129 men went in search of the Northwest Passage in a voyage during which all were lost.

In 1906, Roald Amundsen in *Gjøa* was the first to sail through the Northwest Passage. British, US, Russian and Scandinavian explorers began to vie with each other in efforts to discover new land in the Arctic region and succeeded in charting new areas and discovering new information on geology, climate and resources.

By the turn of the nineteenth/twentieth century, all attempts to reach the North Pole had met with failure, although several expeditions had reached high latitudes in the north. It was into this era of attempts to reach the North Pole that SS *Terra Nova* was chosen to rescue members of the Fiala–Ziegler Arctic expedition.

Captain Johan Kjeld Kjeldsen took command of *Terra Nova* in January 1905. He had a reputation for outstanding seamanship and knowledge of the Arctic Ocean. Johan Kjeld Kjeldsen was born at Bakkejord, Tromsøysund, Norway on 27 July 1840, the son of a blacksmith. He died in 1909, aged 69 years old after

Captain Kjeld Kjeldsen (1840–1909), who took up appointment in 1905 to command *Terra Nova* to relieve the Fiala–Ziegler Arctic Expedition, 1903–05. (Heiddi Siebel/Ivor Stokkeland)

a distinguished career at sea as a sealer, ice-pilot and master of many ships, taking part in a number of scientific, exploratory and sporting expeditions in Arctic waters. He went to sea as a boy and by 1856 he had already sailed to the north polar regions and subsequently gained marine qualifications to captain.

By 1905, Kjeldsen was already a veteran of many Arctic voyages and rescues. He had been appointed to captain *Frithjof* and had taken part in William Ziegler's sponsored first and unsuccessful attempt to reach the North Pole from Franz Joseph Land.

It was the crushing by ice and sinking of the *America* that resulted in Captain Kjeldsen, again in command of the *Frithjof*, firstly attempting an unsuccessful rescue of the Fiala–Ziegler Arctic Expedition in 1904. Now followed the successful rescue of the expedition in 1905, in command of the SS *Terra Nova*.

*Relief of the Fiala–Ziegler Arctic Expedition, 1905*
Of his command of *Terra Nova*, Captain Kjeldsen wrote:

> The ship was bought in London in January [1905] and I as the ship's leader, got telegraphed orders to meet Mr. Champ, Ziegler's secretary, to put the ship in order, among other things. I then travelled to Tromso, where a full crew was hired for the journey. There were 24 men, whereof 15 accompanied me to London to make ready the ship and the entire expedition. The ship departed from London, May, 16th, went to sea, May 18th and arrived in Bergen, May 23rd after a stormy voyage over the North Sea. Departed from Bergen, May 25th and arrived in Tromso, May 30th.

An unimportant difference in dates appears here. Recorded in the diary of Mr F.L. Andrassen, 1st mate of the *Belgica*, is an account of *Terra Nova*'s presence in Tromsø harbour:

*Terra Nova*'s bell on the forecastle, 1905. (Scott Polar Research Institute)

Captain and crew of *Terra Nova*, relief of Fiala–Ziegler Arctic Expedition, 1903–05. Centre row (left to right): Captain Harry McKay, David Bruce (broker), Captain Kjeldsen, William S. Champ (with dog). Also identified: Johan Hagerup, Einar Rotvold, Mr Haugen and Morten Olaisen. (Courtesy Heddi Siebel/Ivar Stokkeland)

We arrived Tromso 29th May (1905) in the evening 9 o'clock and anchored in the harbour, *Terra Nova* before us. She is going to Franz Joseph Land to investigate Ziegler Polar Expedition. Mr. Camp will sail on board – Ziegler is dead and therefore is *Terra Nova*'s flag at 'half-mast all days'.

Captain Kjeldsen continues:

During the stay in Bergen a telegram arrived that Mr. Ziegler had died. This was very hard to accept, but orders arrived that the expedition should continue its journey. After a few changes to the ship were made in Tromso, and provisions, coal and full crew taken aboard, Mr. Champ also came aboard here, after having put the other expedition in order which was supposed to go over to Greenland and look for the depots which were set up there in 1901, together with finding out if any of the polar expedition's participants had arrived there. [A probable reference to the duties of the *Belgica* in the Baldwin–Ziegler Expedition.]

*Terra Nova* departed from Tromso, 14th June at 9.00 p.m. The crew aboard was 21 men in addition to Mr. Champ and Dr. Munth. Everything well aboard.

Ice in sight 9.00 p.m. June 19th 75-1/2 degrees North Lat and 30 degrees 34 East Long. Greenwich. From here the course is set more easterly through the ice, which in part was quite tightly packed. Weather and wind quite variable. Because of very thick fog such that one could not see to come through the ice, we held it back to 75 degrees 40' North Lat. and 44 degrees 45' East Long. Greenwich, during slow penetration of the ice, when one went at full speed from the engine.

One continued thusly until one could come no further north, when the ice became steadily thicker, with large ice floes the further into the polar ice one comes. Several times we were stuck, so nothing could be done to progress. Once in a while we shot a bear, the only diversion we had in our frozen captivity.

The fourth of July, the Americans' big Independence Day, flags on all the tops and a party on board. A bear with its young came out of curiosity and wanted to pay us a visit. The mother was shot and the cub taken aboard alive. It was really naughty young lady of a bear, who didn't want to be obedient, no matter how good we wanted to be to it. Several soundings were taken from 125 to over 200 fathoms and the bottom was such, Whitefish were seen, and one was shot by our harpooner.

They continued north in improving weather and on 29 July the fog lifted and they found themselves approximately 2 miles from land. Captain Kjeldsen continues:

> The course is set for Cape Flora along land, with sharp lookout from deck and crow's nest. From the crow's nest is shouted, 'I think I see people and tents up on that Cape ahead to starboard!'. All eyes are directed there and soon after it proved to be so, that there were people. Here we came across 6 men with 20 dogs lying on a walrus catch and on the lookout for the rescue ship. The mood was good before, it became even better now. Now it was only a matter of getting to hear how their expedition had gone.

*Terra Nova* arrived at Cape Dillon on 29 July 1905.

## Return from a Successful Arctic Mission

Lookouts had been posted at Cape Dillon, but they were becoming anxious. It was now late July and they were depressed at the thought that they would be left there for yet another winter. Fog made false sightings frequent. Then suddenly, it happened ... Fiala recorded:

> Sunday, 30th July [1905] Red Letter Day for all the boys ... At 1.30 p.m. just as I finished my dinner, I looked out of the window and saw Long making frantic gestures ... I could make out the words, 'SHIP! SHIP! SHIP!' Well I guess we were all rather excited and could scarcely control our feelings. I took my glass and from the top of the house could make out a good-sized steamer pulling up, just emerging from the fog and heading right toward Cape Flora, having evidently come from Camp Ziegler way.
> After making out the colors flying at all three mastheads and spanker gaff, I told the men to get all ready to leave and when the ship did tie up in front of the camp I could make out her name, 'Terra Nova' and knew her for a fine vessel of her class, better than the 'America' and a little larger. The 'Terra Nova' was a glorious sight as she materialised out of the mist, her form glistening in the sunshine as the fog lifted.
> I never experienced such a sense of loss as I did on beholding the splended vessel; never realised so keenly our shipwrecked condition. A boat brought in Mr. Champ and Dr. Mount and as Mr. Champ said, ''twas the happiest day

Arrival of *Terra Nova* at Cape Dillon, Franz Joseph Land. (Courtesy Heddi Siebel)

of his life ... everything was all ready to put on board and we did not detain the stmr [steamer] any as they did want to get back to Ziegler [camp] soon as possible to get the rest of the expedition.

Soon all were aboard *Terra Nova*. For the first time in well over a year, the entire expedition was together. Only Myhre, the Norwegian who had died in May 1904, was missing. His body would remain forever at the northern tip of the archipelago, 525 miles from the North Pole they had never reached. Among the letters in the mail waiting for them aboard the relief ship was one addressed to Myhre which informed him that his wife had died. By an uncanny coincidence, she died the same day he did, neither aware of the other's death.

The men now enjoyed the comfort of warm cabins, hot baths, good food, wine, clean clothes, cigarettes, cigars and letters from home. The men at Cape Flora had packed in such frantic haste that they left some items behind. *Terra Nova* headed back to get them. Even Captain Coffin had forgotten things:

> August 1 Steamed in and put on small board for the beach ... Opened the house and got my barometer, sewing gear, toothbrush and pair of new moccasins, all overlooked. Most of the men found some things they wanted. Nailed up the door and left everything OK ... Started for Norway at 3.30 p.m.

The trip to Norway was sheer pleasure except for a financial settlement to each of the expedition members. Fiala called for all to assemble in the main cabin. The men had already been informed of the sad news that Mr Ziegler had died two months previously, but Ziegler had set aside a large sum of money for distribution which he had planned as a happy and grateful celebration. The amount each was to receive was to be based on individual contribution to the expedition in terms of loyalty and value of services rendered, an assessment to be decided by the expedition leader, Fiala. It was only to be expected that the meeting was reduced to unpleasantness when Fiala offended some of his men with his allocations of the bounty, and the reasons given remain a mystery.

Of the return south, Fiala wrote:

Under Captain Kjeldsen the splendid 'Terra Nova' forced her way through the icy Barents Sea. It is always easier to leave the pack ice than to enter it, to go south than to go north, and on Sunday, 6th August, we entered the open sea and felt for the first time the motion of the waves.

Captain Kjeldsen continues:

On the 6th August 1905, Lat. 77 degrees 33' North, Long. 53 degrees 47' East. In the afternoon the ice began to be more and more separated and open sea finally at Lat. 77 degrees 12' north, Long. 53 degrees East. Now the log is set, together with all sails, and course is set for Norway.

Wednesday morning at 5.00 a.m. land in sight at Vardo. The current has sent us eastward while we were stuck in the ice the 5th and 6th, much more than we had reckoned. Our course now goes along the coast under sight and sound of land. Eleven o'clock in the evening we stopped in Honningsvag in order to send telegrams for several hundred kroner, the largest amount, said the telegraphist, that he had taken at one time – even though(t) he was the same man who greeted 'Fram' when it came to Skjervo.

When we came out of the ice to the open water, 40–50 dogs were shot and thrown overboard, some were spared, because they had done so much for their masters. Four of the dogs came all the way over to New York and were given to the zoo together with the 3 bear cubs and the live fox and the polar gulls, which I later saw there, all in the best of shape.

Thursday evening, the 10th August, at 10.00 p.m. we arrived in Tromso. All is well, some of the expedition's participants left us here. Saturday evening, the 12th August, we departed from Tromso for Bergen, where we arrived 11.30 p.m. the 16th August. Here all of the expedition's participants were

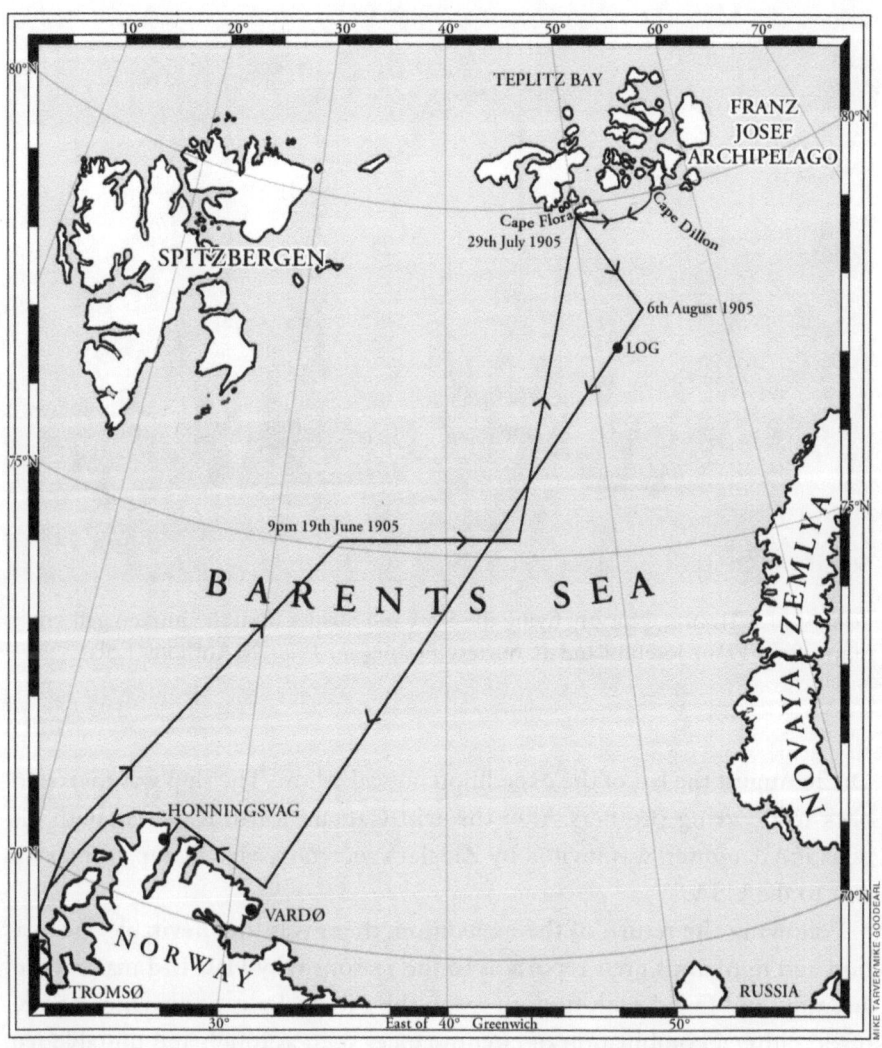

1905 relief voyage of *Terra Nova* from Tromsø, Norway, to Cape Flora and Cape Dillon, Franz Joseph Land. (Chart compiled by author)

cleared by American consul and several left us here, but some accompanied us over to London. Mr. A. Fiala boarded the England boat at Stavanger. Now we had a storm with headwinds and lay at anchor in Hoivarde until the 21st August. When the weather got better, we went to sea and a pilot took us up the Thames to London. On the 25th August, the 'Terra Nova' was towed in to West India Dock at 10.00 p.m. All is well aboard.

*Terra Nova* at Tromsø, Norway, flying the Stars and Stripes from the mizzen gaff on return from Franz Joseph Land. (Courtesy *Fighting the Polar Ice*, Anthony Fiala)

On 26 August the last of the expedition moved ashore. The ship was then sold back to Bowring Brothers. After this trip, Captain Johan Kjeldsen, with his wife and daughter, was invited by Ziegler's secretary, Mr Champ, to make a visit to the USA.

Following the return of the expedition, there was the inevitable speculation and numerous press reports as to the reasons why Fiala had made three attempts and failed each time to reach the North Pole. Explanations were made, citing a combination of circumstances such as rough and unstable ice conditions and dangerous openings of clear water which could not be negotiated. Because conditions were beyond the equipment and capabilities of the expedition, the furthest north reached was Lat. 82° N. Anthony Fiala told his story of the expedition in *Fighting the Polar Ice*, published in 1907 and a volume of scientific reports was also published.

They had not been further north than Captain Coffin had reached in the expedition ship, *America*. At least Captain Edwin Coffin was happy with his navigational achievement to Lat. 82°14' N, even though he had lost his ship.

Captain Edwin Coffin's expedition papers are in the archives of the Martha's Vineyard Historical Society, at his home town of Edgartown, Massachusetts, USA. In the story 'Shipwrecked 525 Miles from the North Pole', serialised in the society's magazine, *Dukes County Intelligencer*, Arthur R. Railton tells Captain Coffin's story of the expedition taken from his personal papers and notes. However, Captain Coffin apparently did not write another word in his journal after the arrival of the relief ship, or if he did, nothing is included in his archived papers.

In a tribute to Captain Kjeldsen, the expedition manager, William S. Champ wrote:

> In conclusion permit me to introduce to the readers of this narrative, Captain J. Kjeldsen of Tromso, Norway, a true action hero, the man who safely navigated the 'Terra Nova' which effected a timely rescue of the members of the Ziegler Expedition. To him and his faithful Norwegian officers and crew, the writer feelingly tenders this acknowledgement and publicly expresses the heartfelt appreciation of the rescue.

William S. Champ
New York, Aug. 14th, 1906

## Return to Newfoundland for Sealing Duties

A list of Bowring Company sealing catches attributed to *Terra Nova* show that the ship soon departed London and returned to St John's, Newfoundland. In 1906, she was under the command of Captain Abram Kean and once again operating from Newfoundland. She is shown in that year to have taken 16,627 seals. In 1907 she took 18,785 seals and in 1908, 13,962 seals, again under the command of Abram Kean.

In 1909, she was under the command of Captain Edward Bishop and brought in 15,468 seals. Her reputation as a polar expedition ship, both in the Antarctic and now the Arctic was firmly established. In the year 1909, she was sold by Bowring Brothers Ltd to the British Expedition Committee for Captain Robert Falcon Scott's 1910 expedition to the Antarctic.

This was to be SS *Terra Nova*'s most famous voyage.

Sealing fleet in the port of St John's Newfoundland, around 1905. (Courtesy Tundra Books, Toronto)

# 5

# HER NAME WILL BE REMEMBERED FOREVER

Expedition Ship for the British Antarctic Expedition, 1910

In 1909 *Terra Nova* was sold by C.T. Bowring & Co. Ltd to be used in her most famous role – to become the expedition ship for the second, most famous but tragic British Antarctic Expedition, 1910, to be led by Captain Robert Falcon Scott RN.

After the success of the British National Antarctic Expedition of 1901–04 in the RRS *Discovery*, Scott returned home to national acclaim. He had been as far south as Lat. 82°11' S, 469 nautical miles from the South Pole. This expedition had made important scientific discoveries. Scott had achieved much success and must have felt that although there were things to be done at home, there was unfinished business in the Antarctic. The desire to return was always present.

It was, however, necessary to pursue his Royal Naval career, seek financial stability and pay attention to family responsibilities. Promoted to captain, he was granted leave and set about writing *The Voyage of the Discovery*, his account of the 1901–04 expedition. He then returned to operational naval duties in command of warships. He married Kathleen Bruce on 2 September 1908.

Correspondence between Scott and a former member of his expedition, Ernest Shackleton, more than kept alive the left over business in the Antarctic, with exchanges over 'territory' and use of the 1901–04 Base Hut in McMurdo Sound. Shackleton, who was in ill health, had returned home from the expedition on the relief ship, *Morning* which had called after the first winter in 1903. A highly motivated and ambitious man himself, Shackleton had been deeply hurt by the indignity of having to return home prematurely and was planning to return to the continent with his own expedition.

When Shackleton returned to Antarctica with the *Nimrod* Expedition in 1907 and set up a Base at Cape Royds, Scott was an anxious observer. Shackleton's achievement in going further south and getting to within 97 nautical miles of the South Pole clearly left Scott believing that he himself must return and was the man destined to take the final prize.

The circumstances surrounding the relationship between Scott and Shackleton, include the premature return home of Shackleton from the 1901–04 Expedition and his subsequent success in returning to lead his own expedition in 1907. On his return in 1909, Shackleton received a knighthood. He had made a greater penetration of the continent beyond Scott's achievement, and had almost succeeded in reaching the South Pole. This was the true stuff of 'one-upmanship'. Scott in returning to 'finish the job' set the scene for further excitement and anticipation.

Many enthusiasts of the heroic age of exploration choose to take sides and make comparisons between these two leaders, concluding that one must be better than the other and that in favouring one, they must be against the other. Perhaps this is all part of the interest and curiosity with the subject which keeps the stories alive with never-ending fascination. A closer study by students of the subject reveals no serious personal animosity between the two men other than jealousy of each other's ambitions, though certainly there was controversial correspondence over the use of the hut at McMurdo Sound.

It is evident that they were two very different types of leaders; men of contrasting personalities, both were professionally and socially ambitious in different ways. Scott was the Establishment man with socially accepted connections and expectations, a Royal Navy trained officer with the initial sponsorship, support and appointment by the Royal Society and Royal Geographical Society in 1901. He was a more serious-thinking man, used to the discipline of a military service and the loneliness of command, with a resulting sense of national responsibility. He was more used to formality and the methodical organisation which would be expected of him, and respected and understood by expedition members, who were either still-serving officers and ratings in the Royal Navy or past members.

A long-term study of Scott from works recorded by those who knew, lived and worked alongside him, does not reveal him to be an arrogant and bungling leader. Certainly, the loneliness of command affects people in different ways, sending some within themselves, bringing with it moods, and it could not be suggested that no mistakes were ever made by him. Scott is cynically portrayed

```
Telephone No.: 1490 GERRARD.
              BRITISH ANTARCTIC EXPEDITION, 1910.
                    Advisory Committee:
Major LEONARD DARWIN, R.E. (President, Royal Geographical Society).        36 & 38, VICTORIA STREET,
The Right Hon. LORD STRATHCONA, G.C.M.G., &c.                                        S.W.
Sir CLEMENTS R. MARKHAM, K.C.B., F.R.S.
The Right Hon. Sir GEORGE D. TAUBMAN GOLDIE, P.C., K.C.M.G., &c.
The Right Hon. VISCOUNT GOSCHEN.
The Right Hon. LORD HOWARD DE WALDEN.
Sir EDGAR SPEYER, Bart. (Treasurer).
                         Bankers :
         Messrs. COCKS, BIDDULPH & CO., 43, Charing Cross, S.W.
                        Auditors :
         Messrs. JAMES FRASER & SONS, Chartered Accountants.
```

as a bungler by Roland Huntford in his story with many absurd conclusions, *Scott and Amundsen*, which is, in the opinion of experts, a flawed work with a preplanned agenda of tabloid-style presentation offering up the debunking of a national figure.

Ernest Shackleton, as a personality, seems to have had more of the style of a privateer with a cavalier approach, laced with over-optimism. He was his own man with a large and charismatic personality and was popular and engaging, less an Establishment man and without formal rank in a military service, but ambitiously seeking to be accepted within the Establishment. A Merchant Navy officer, and used to command, he was confidently at home with everybody. He had a free, easy, open and energetic personality; a quality which men will generally like to follow.

Put both personalities together and it could be said that while Shackleton would do the talking, Scott would do the thinking. People of such personalities will often grate on one another when working together. Yet, these arguments can be debated forever and turned on their heads, for one might say that of the two, Shackleton was the more reckless with his over-optimistic expectations. This made him a greater risktaker, in many ways, particularly in terms of his planning and preparation, as shown in the outcome of his well-known Imperial Trans-Antarctic Expedition of 1914–17, with the ships, *Endurance* and *Aurora*.

Yet, it was Scott who admitted to taking risks which, in his own words, 'came out against us'. But exploration of unknown regions has to be about taking risks to their very limit. As a subject of endless debate and comparison of personalities, the heroic age of polar exploration continues to be a subject of fascinating intrigue.

Now, with the 'prize for the taking', Scott saw that he must pursue it and return to the Antarctic. He clearly saw himself as the obvious British candidate to take that opportunity, for explorers of other nations were also seeking the honour of being first at the South Pole. At a dinner held at the Savage Club,

London, to welcome Ernest Shackleton back on 19 June 1909, Scott declared, '… the Pole must be discovered by an Englishman'.

On 13 September 1909, Scott published his plans for the British Antarctic Expedition of 1910. An office was taken at 36 & 38 Victoria Street, London, SW. Dr Edward Wilson was appointed as head of the scientific staff. Mr Francis R.H. Drake, assistant paymaster RN, was appointed secretary and Sir Edward Speyer, a London financier consented to be the treasurer. Now, an expedition ship had to be found.

## Fitting Out in West India Dock, London

The choice of the ship was made on 22 September 1909. The RRS *Discovery* was not available. After the 1901–04 expedition, she had been taken over by the Admiralty and then sold to the Hudson Bay Company for conveying cargo, and the company would not release her. The obvious next choice was now SS *Terra Nova*, the largest and strongest of the Scottish whalers, which by now had already earned a reputation in both polar regions.

*Terra Nova*'s owners, C.T. Bowring & Co. of Liverpool and St John's, Newfoundland, agreed to release the ship for the expedition before a tenth of the agreed purchase cost of £12,500 had been raised. As with all

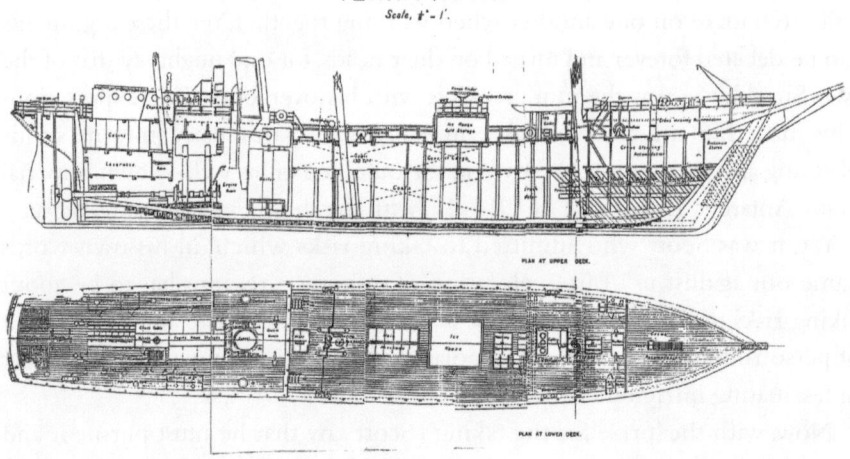

Hull plans for the expedition refit. Supplied by David Bruce & Co. Ltd, London. (Courtesy Scott Polar Research Institute)

Awaiting refit at West India Dock, London – 'poor little ship, she looked so dirty and uncared for'. (Courtesy *Western Mail*)

expeditions, raising money was the hardest part of the enterprise. Scott wrote on 25 November 1909, 'Made an agreement with Bowrings to pay only £5,000 on the 8th, don't know how I shall manage even that but must see Speyer next week.' By the spring of 1910, £10,000 had been raised from the public and the government had added a grant of £20,000.

*Terra Nova* was surveyed in Newfoundland and ownership was handed over to the expedition on 8 November 1909, at the West India Dock, London. The purchase had been arranged by brokers David Bruce & Sons, who subscribed the amount of their commission to the expedition and supplied hull plans for the refit. The owners, C.T. Bowring & Co. also subscribed £500 and greatly assisted in raising money in Liverpool.

Scott appointed Lieutenant E.R.G.R. 'Teddy' Evans RN as second in command of the expedition, and as commander-elect of the ship, and mainly left to him the choice of the officers and crew. Dr Wilson, assisted by Captain Scott, selected the scientific staff. *Terra Nova* was docked by the Glengall Ironworks Co., to whom strengthening and fitting out was contracted according to specifications required by the expedition.

'Teddy' Evans described much of the refit to the ship in South with Scott:

> We were somewhat limited in our choice of ship suitable for the work contemplated ... the 'Terra Nova' had one great defect, she was not economic in the matter of coal consumption. She was the largest and strongest of the old Scottish whalers, had proved herself in the Antarctic pack ice and acquitted herself magnificently in the northern ice-fields and in whaling voyages extending over a period of twenty years. In spite of her age she had considerable power for a vessel of that type ... I shall never forget the day I first visited the 'Terra Nova' in the West India Docks, she looked so small and out of place surrounded by great liners and cargo-carrying ships, but I loved her from the day I saw her there, because she was to be my first command. Poor little ship, she looked so dirty and uncared for and yet her name will be remembered for ever in the story of the sea
>
> [...] I often blushed when Admirals came down to see our ship, she was so very dirty. To begin with, her hold contained large blubber tanks, the stench of whale oil and seal blubber being overpowering and remarks of those who insisted on going all over the ship need not be set down here. However, the blubber tanks were withdrawn, the hold spaces got the thorough cleansing and white washing that they so badly needed. The bilges were washed out, the ship disinfected fore and aft and a gang of men employed for some time to sweeten her up.

Examining the standard compass fitted to the icehouse roof (left to right): Lieutenant Evans, Lieutenant Campbell, Lieutenant Pennell. (Courtesy *Western Mail*)

[…] The hull was reinforced from the interior with oak beams and she was reinforced considerably at the bow and stern with exterior steel plating. She was returned to her original rig as a barque, and on the upper deck a large well insulated ice-house was erected. This was to hold 150 carcases of frozen mutton. Owing to its position, free from the vicinity of iron and with a good all-round view; the top of the ice-house was selected for mounting the Standard Compass and the sounding machine by Kelvin & James White, also, the Lloyd Creak pedestal for magnetic observations and the range-finder by Barr & Stroud.

The galley was almost re-built and a new stove put in. The forecastle was comfortably fitted up with mess-tables and lockers. A lamp room was built and paraffin tanks installed to hold 200 gallons for lighting purposes. Storerooms, instrument room and chronometer rooms were added.

The greatest alteration was made in the saloon, which was fitted with bunks to accommodate twenty-four officers and men. This was scarcely luxurious accommodation but it was always kept clean and the ventilation was good. Then a nice little mess was built for the warrant officers, of whom there were to be six. For the warrant officers and seamen, hammock spaces or bunks were provided.

Two large magazines and a clothing store were constructed in the between decks. These particular spaces were zinc lined to prevent damp creeping in. It was found necessary to put a new mizzen mast into the ship, but on the whole she required alteration rather than repair.

All the blubber tanks were withdrawn and the hold spaces thoroughly cleaned and whitewashed. A good chart-house was built over the wardroom and a large covered chart table fitted up on the bridge. The Glengall Co. were most anxious to meet us in everything and to push the alterations forward, and their work was efficient and not expensive. The original date for sailing was for 1st August, but by the united efforts of all concerned with the fitting out and stowing of the ship, the time apportioned for preparing the vessel was halved and she was ready for 1st June.

The ship had to be provisioned and stored for her long voyage and here again lists had to be prepared to meet every contingency. There were boatswain's stores, wire hawsers, canvas for sail making, carpenter's stores, and cabin and domestic gear to be provided. The engineers had to purchase their stores along with blacksmith's tools. There were fireworks for signalling, whale boats and whaling gear, flags, logs, paint and tar and a multitude of necessities to be thought of, selected – and not paid for if they could help it. The expedition had many friends, among them members of the RNVR (Royal Naval Volunteer Reserve), working on board and who gave of their services at weekends for nothing.

An invaluable collection of polar literature, Antarctic and Arctic, was made up for the expedition by Admirals Sir Lewis Beaumont GCB and Sir Albert Markham KCB and the beautiful library in miniature was presented by Mr Reginald Smith.*

Numerous companies with well-known names of the day, many of which are still in business, provided foodstuffs, tobacco, chocolate and clothing. A piano was provided by the Broadwood Co. and two gramophones by the Gramophone Co.

The ship was also equipped with ice-saws, anchors, picks and shovels, hides for soles of boots, instruments of all descriptions for various scientific purposes, lamp and lighting gear, mathematical tables, the library, oils and mineral grease, a colossal photographic outfit, stationery in huge quantities, sledging outfits, harnesses and leather goods for ponies and dogs and motor accessories for the transports. There were telescopes, binoculars, fishing gear and sewing

---

\* The wooden bookcase is now at the Scott Polar Research Institute.

Discussing the refit. Left to right: Lieutenant Bowers, Lieutenant Evans, Lieutenant Campbell. (Courtesy *Western Mail*)

gear, including sewing machines presented by the Singer Co. The Welsh Tin Plate & Metal Stamping Co. of Llanelli, provided for free the majority of the cutlery, cooking apparatus and mess traps. Lists were trawled for what might be forgotten as Captain Scott had once complained that they had forgotten to take shaving soap – but it was produced!

The officers arrived to take up their duties. Lieutenant Campbell as first officer, Lieutenant Pennell as navigating officer and Lieutenant Rennick. Lieutenant Bowers, who came from the Royal Indian Marine, was stores officer but then promptly fell down the main hatch.

The incident was reported to Lieutenant Evans, who said, 'What a silly ass, after coming all the way from Bombay, falling down the main hatch seems a bit careless to say the least'. When it was reported that Bowers had not hurt himself, Evans responded, 'What a splendid fellow!' Bowers had in fact fallen 19ft without injuring himself. He would prove to be one of the toughest men on the expedition.

In his account of the expedition, *South with Scott*, Evans described the colours flown by the ship:

> Most polar expeditions sail under the burgee of some yacht club or other. We were ambitious to fly the White Ensign and to enable this to be done Scott was elected a member of the Royal Yacht Squadron. He then went on to bemoan the fact that it cost us £100.00 which we could ill afford.

If they sailed as a merchant vessel, they would be obliged to fly the Red Ensign. However, with *Terra Nova* registered as a yacht they were able to avoid those Board of Trade Officials who might have declared the ship not a well-found merchant ship within the meaning of the act. Evans went on, 'Having avoided the scrutiny of the efficient and the official, we painted out our Plimsoll mark with tongue in cheek and eyelid drooped, and, this done, took our stores aboard and packed them pretty tight.'

When the day came to depart from London, Captain Scott called all hands aft and thanked them for what they had done. The work had been completed, transforming the grimy vessel of which they had taken delivery in the previous November. It was now 1 June 1910.

Lady Bridgeman, wife of the First Sea Lord, and Lady Markham hoisted the White Ensign and the burgee of the Royal Yacht Squadron an hour or so before sailing. At 4.45 p.m. the visitors were warned off the ship and a quarter of an hour later, *Terra Nova* slipped from the wharf at the West India Dock to Greenhithe where she anchored off the mercantile training ship *Worcester* giving a chance for the cadets to view the expedition ship.

On 3 June 1910, *Terra Nova* arrived at Spithead where she was boarded by Captain Chetwynd, Superintendent of Compasses at the Admiralty, who swung the ship and adjusted the compasses. Captain Scott rejoined the ship on 4 June and paid a visit with his yacht to the Royal Yacht Squadron at Cowes. On 6 June, a series of magnetic observations were completed in the Solent. Later, the officers were entertained by Captain Mark Kerr aboard HMS *Invincible*, where they took advantage of the hospitality and cleared out a quantity of canvas and boatswain's stores, including a 3½in hawser. Good use of their spoil was made on *Terra Nova*. (HMS *Invincible* was sunk at the Battle of Jutland in 1916.)

On 7 June, a farewell fleet bade them goodbye from the Solent and via Weymouth where Scott rejoined the ship, they set course for the Bristol Channel and Cardiff. They arrived off Cardiff on the evening of Thursday, 9 June 1910, in preparation to enter the Welsh port the following day, where they spent six days before the final departure from the UK for the south.

City of Cardiff Civic Centre, around 1905. (*A Guide to Cardiff, City & Port* by John Ballinger)

## Preparations and the British Departure from Cardiff

*Terra Nova* arrived off Cardiff earlier than expected and a welcoming party led by the Lord Mayor was hastily put together, which sped out on the tug *Mercedes* to make its initial welcome. *Terra Nova* lay overnight at anchor in Cardiff Roads, awaiting a pilot and tugs to take her in tow on the following morning, when a more formal welcome had been arranged. She was about to enter one of the great steamer ports of the world.

The notion that *Terra Nova* merely 'called at Cardiff to take on coal' is very much an understatement, and it is worth examining the visit in more detail. The port of Cardiff and her neighbouring ports of Penarth and Barry were at their peak 100 years ago as the greatest coal exporting ports in a world economy driven by coal and steam. Such was the loading activity and capability in those days that a vessel of *Terra Nova*'s size could have taken on her coal requirements in an hour or two, or even less, then be turned around and cast off back to sea. Such was the frenzy of activity in those 'heady days' when ships had to await berthing room in the docks.

In fact, the six day visit to the city of Cardiff by *Terra Nova* was the culmination of crucial final preparations before departure from Britain. It was an expedition desperately in need of every generous offer it could find, being still short of money to pay its members.

The early part of the nineteenth century saw Cardiff as little more than a small fishing and trading port and as the Industrial Revolution firmly took hold, demands for coal increased both nationally and for export. Docking facilities became inadequate as demand outstripped the river wharfs and the small canal dock already in use. From 1839 to beyond the turn of the century, new docks were built and with export activity increasing, more new docks were built at the neighbouring towns of Penarth and Barry.

A census of 1851 shows there were just 20,000 Cardiff inhabitants. As the handling capacity of the new docks increased, trade multiplied, and in the record year of 1913, 13.7 million tons of cargo were exported, 10.5 million tons in coal alone. As trade developed, the growing port became a 'honeypot' for ship owners, ship builders, mine owners and all the ancillary trades of ship's chandlers', with work for dockers and seamen.

In 1905, the Borough of Cardiff was granted city status. The handsome civic centre buildings in Portland stone were completed, enhancing the face of the city and giving it a civic centre that was said to rival that of any city in the world.

Above all, it must be said that the city of Cardiff owes its prosperity to the miners and their families of its valley hinterland, for it was the men at the coal face who bent their backs to produce the 'black diamonds' and who saw coal carried by rail to the docks, to produce the wealth which brought about a seaport known throughout the world. Little wonder, then, that the docks businessmen in those prosperous days saw the potential of backing a famous expedition to the unknown end of the world in the south as a way of promoting their businesses and their city.

A comprehensive study of the Cardiff connection with the British Antarctic Expedition of 1910 was made in 1984 by Dr Anthony M. Johnson of University College, Cardiff. His detailed analysis of the contribution made by the docks businessmen and their associates to the finance, equipment, funding and stores for the expedition is shown in great detail in his study, *Scott of the Antarctic and Cardiff* (now in its third edition).

The idea of supporting an Antarctic expedition came about after a contact was made in the city by Lieutenant E.R.G.R. Evans, who had in mind the idea of promoting an expedition of his own. Evans had been second officer in the relief ship *Morning* when it made its relief voyages to the *Discovery* expedition in 1903 and 1904. There are few whose imagination could fail to be fired by the beauty and awesomeness of the Antarctic continent, and Evans, an ambitious young naval officer, would have listened to the adventures of expedition members returning aboard *Morning*, who included Ernest Shackleton, who had accompanied Scott and Wilson toward their 'farthest south'.

Evans had, through sources in London, already begun to make plans for an Antarctic expedition led by himself and based on Cardiff. This news had been communicated to the editor of the *Western Mail*, William E. Davies, which had prompted him to travel up to the capital from Cardiff in pursuit of the story. Evans disclosed his plans to the editor over dinner and Davies agreed that such an enterprise based on Cardiff was worth pursuing. Davies was already in a position to interest the commercial community of the city because of his existing connections.

News that Evans was serious in his ambitions caused concern that he would emerge as a serious rival to Scott's plans, although it seems that neither was aware of the other's intentions. Contacts were made, eventually resulting in Sir Clements Markham bringing Scott and Evans together, with the result that Evans withdrew his plans and agreed to join Scott, as leader, with himself as deputy.

Now, with commercial support in the air and the leadership settled, the outcome of these earlier meetings eventually brought about more backing and, most importantly, the commercial backing of Cardiff's business community. This was enthusiastically led by Daniel Radcliffe, of the ship and coal owners, Evan Thomas, Radcliffe & Co., who were the port's largest ship-owning company with a fleet of twenty-six ships.

Evans's visits to Cardiff were to prove fruitful, and with the name Evans, although not himself born in Wales, his reported Welsh family origins were readily accepted and all was heartily absorbed into the intense enthusiasm to support the expedition.

Dr Anthony Johnson, in his analysis of the Cardiff connection, continues his detailed story:

On the morning of Friday, 10th June 1910, the Lord Mayor's welcoming party went out in Mr Edmund Handcock's tug, 'Falcon' and boarded the 'Terra Nova' waiting at anchor in Cardiff Roads ready to enter port. At 8.45 a.m. with the welcoming party aboard, 'Terra Nova' weighed anchor and with channel pilot, Mr. Thomas Jenkins in charge, was taken in tow by Mr. Handcock's tugs, 'Falcon' and 'Plumgarth'. Evans in *South with Scott* describes the arrival … 'We were welcomed by the citizens of the great Welsh seaport with enthusiasm. Free docking, free coal, defects made good for nothing, an office and staff placed at our disposal, in fact everything was done with an open-hearted generosity.'

A busy scene at the Bute East Dock, Cardiff, Saturday, 11 June 1910. (Courtesy *Western Mail*)

*Terra Nova* was taken to the Roath Dock and berthed at the Crown Patent Fuel Company wharf, where the offices were dressed overall for the occasion. Here, *Terra Nova* took on 300 tons of Crown Patent fuel, a patented name for compressed briquettes of coal and bitumen. This was a particularly useful item of fuel as it was easy to stow, took up less room than ordinary coal, retained its heat for long periods and proved useful later at base camp in providing temporary walls as partitions to shelter the animals.

The following morning, Saturday, 11 June, *Terra Nova* moved to the East Bute Dock to take on board 100 tons of steam coal, 50 tons of which were the gift of Ynyshire Colliery Company and 50 tons from Messrs. Insole and Sons. The Vacuum Oil Company made a gift of 500 gallons of engine and lamp oil. Many other items of equipment and stores were taken aboard throughout the weekend and, during the six day stay, Mr William Jones, manager of the Channel Dry Dock Company, put his office, telephone and all other facilities at the disposal of the expedition.

When the loading of the 400 tons of coal and fuel was completed it brought with it a problem. The ship had now settled lower in the water and seams in the hull which had been above the water line now leaked. The ship was dry docked and efforts were made to fill the seams by caulking and cementing. The condition of the ship and its leaks became the subject of much press speculation. Certainly, considerable time was going to have to be spent pumping out the ship as she went south.

The officers and crew spent a busy weekend attending the many social events that had been arranged for them. Some expedition members joined the ship at Cardiff, amongst them Dr Wilson. On 13 June, he and his wife had tea in London with Sir Clements and Lady Markham and that evening, the Wilsons caught a train to Cheltenham. The journey was spent busily writing a report on grouse plumage, a project on which he had become involved after been appointed by the Board of Agriculture to look into a disease that was ravaging grouse stocks on Britain's moors.

After a busy day at home, the Grouse Commission papers were completed and put in the post. That evening, 14 June, the Wilsons took the train to Cardiff to join the expedition. Officers, crew and scientists of the expedition had all assembled and departure day was to be Wednesday, 15 June 1910.

A farewell dinner for Captain Scott, his officers and scientists, was held at the Royal Hotel in the town centre on Monday, 13 June. Here, they were entertained by the business community of the city. Some 200 gentlemen were present for the ten-course dinner priced at 7s 6d, which included fillets of beef 'Terra Nova', soufflé 'Captain Scott' and 'South Pole' ice pudding. They were entertained by the F.G. Roberts String Band, who discreetly played a selection of numbers including 'Hero of the South' and 'Farewell of the Gladiators'. All was washed down with Veuve Cliquot, Dry 1900 and Gautier Frères Very Fine Cognac (twenty years old).

From the dinner table came more pleas for donations and Daniel Radcliffe began the acts of generosity by making a personal donation of £500. Within a short time, it was announced that donations that evening had amounted to £1,000 and that Cardiff in donations alone had exceeded Manchester in its generosity. The total subscriptions and donations eventually came to £2,500 from Cardiff, in addition to the assistance given with equipment, fuel and marine services. During the evening, Scott expressed the view to his principal supporters that in recognition of its commitment and generosity, the city of

Stern view of *Terra Nova* in the Roath Bute Dock, Cardiff. (*South Wales Daily News*)

Cardiff should be designated as *Terra Nova*'s home port in the UK and would be the first port to which the expedition would return.

The crew members attended a function at the nearby Barry's Hotel with the menu priced at 2s 6d. At the end of the evening, all were together at the Royal Hotel for a raucous finale, where it is said that many exceeded themselves! This, after all, was to be their final farewell in the UK.

The following day the *Western Mail* correspondent who had been present at the dinner reported:

> I have never seen anything approaching the enthusiasm at last night's dinner. Even the meanest of us was stirred and felt like asking whether there was not a spare berth in the ship. The sight of the officers and crew was an inspiration. Such seaman, with faces of bronze and necks that Roman Gladiators would have given the world for, were never gathered before in any ship that ever sailed from a British port.

Many members of the expedition signed a dinner menu card, and among the signatures is the name 'Thos. Feather'. Thomas A. Feather, Petty Officer 1st Class, had been a member of the British National Antarctic Expedition 1901–04 led by Scott and was bosun of *Discovery*. He had taken part in a number of sledging journeys and Scott thought highly of him, recording his valuable contribution to the expedition in *Voyage of the Discovery*, published in 1905. Feather, in enlisting for the expedition on the Crew Agreement for the *Terra Nova*, clearly indicated an enthusiastic intention on his behalf to return south.

However, this was not to be, for in New Zealand Feather was 'discharged with very good conduct', as recorded on the Crew Agreement. The responsibility for affairs aboard ship had been delegated to Lieutenant Evans as captain and it appears he preferred the services of Alfred Cheetham as bosun to those of Feather. Lieutenant Evans would have known the capabilities of Cheetham, as they had served together aboard *Morning* on the relief voyages to the *Discovery* expedition.

Cheetham afterwards had gone on to serve in *Nimrod* with Shackleton prior to being appointed to *Terra Nova* and had now been below the Antarctic Circle four times. Cheetham's background was essentially with the mercantile marine and Lieutenant Evans, too, had begun his seagoing days in the mercantile marine training ship, *Worcester* at Greenhithe. After two years in *Worcester*, Evans won a Queen's Cadetship to the Royal Navy ship *Britannia* at Dartmouth.

Feather, according to the crew list, was 39 years of age. But the crew list also shows Cheetham aged 42 years, so age may not have been a factor. There

Scouts of the 4th Cardiff present their colour to the expedition. (Scout group archives)

is no obvious reason for the preference for Cheetham, other than a mention in *South with Scott*, when Evans records, 'Scott was impressed with my man Cheetham', when referring to the crew. However, Cheetham had by this time been below the Antarctic Circle more times than any other man and as captain of the ship it was left to Evans to make the appointment.

On the evening of Tuesday, 14 June, the civic authorities led by the Lord Mayor held a buffet reception for the expedition members at the City Hall, for which 800 invitations were issued. Musical entertainment was provided by Madame Hughes-Thomas's Royal Welsh Ladies' Choir. Captain Scott, in an address of thanks reiterated that it was the unanimous wish of his officers and crew that should the expedition be successful in its aims then the Port of Cardiff, in preference to any other, should be the first port of call on the return of *Terra Nova* to the UK.

The Board of Trade Agreement and Account of Crew shows the *Terra Nova* as a yacht and her registered managing owner as Robert Falcon Scott, Captain RN, care of David Bruce & Co., 37 Billiter Buildings, London EC. Details required of her crew were entered on the agreement at Poplar, London, dated 30 May 1910 and signed by Edward R.G.R. Evans (master). Of those listed, three were eventually discharged as 'Conduct VG' and three with 'ill-health'. One man deserted the ship at Cardiff.

At 1 p.m. on Wednesday, 15 June 1910, *Terra Nova* passed through the lock gates of the Roath Basin, leaving the Bute Docks on the first stage of her journey south. A claim to be the last man to come ashore from the ship before her departure from Cardiff came many years later from Mr Bill Batstone, a ships' painter who had been working aboard. When he was interviewed, he was well into his nineties. His very last job aboard *Terra Nova* was to paint the crow's nest, which, he recalled with a smile, had been specially supplied and adapted by a well-known Cardiff brewery.

By prior arrangement, Scouts of the 4th Cardiff (St Andrews) Group, founded in October 1908, were invited aboard the ship and presented their Colour to the expedition. The Colour sailed south with the ship. When the SS *Terra Nova*, RYS (Royal Yacht Squadron) left Cardiff docks, Captain Scott, Mrs Scott and the wives of other officers were aboard, together with the Lord Mayor's party and Daniel Radcliffe.

This was the start of *Terra Nova*'s most famous role as a polar exploration ship. The story of the ship's role in Scott's last expedition is recorded for history in letters, diaries and written accounts by expedition members. Of the departure from Cardiff, Dr Wilson wrote in his diary:

> The send off from Cardiff was very enthusiastic, enormous crowds having collected at every available spot to cheer and fire guns and detonators and to make a perfectly hideous din with steam sirens and hooters, of which Cardiff seems to possess an infinite number.

Captain Oates (of landed gentry) was less complimentary and seemed to detest farewells, writing home to his mother, he recorded, 'The Mayor and his crowd came on board and I never saw such a mob, they are all Labour Socialists … the only gentleman I have seen come aboard is the telephone operator.'

To add to the excitement, one member of the crew fell overboard but was safely recovered. (This is a press report and examination of the ship's log at the Scott Polar Research Institute does not show an entry to this effect, if it occurred.)

In his story written seventy years later, Roland Huntford, in *Scott and Amundsen*, describes the scene in his perverse account of the story as if he were there himself. He writes, 'Here at Cardiff was the authentic roaring of the plebs.' Does he mean that all those who waved goodbye to the expedition must also have been plebs?

The departure from Bute Docks, Cardiff on Wednesday, 15 June 1910. (Courtesy *Western Mail*)

*Terra Nova* sailed from Cardiff as a yacht. As SS *Terra Nova*, RYS, she flew the city of Cardiff coat of arms from the foremast and the Welsh dragon at the mizzen mast. Captain Scott was aboard, so she also sailed under the White Ensign, with the burgee at the head of the mainmast signifying her registration as a yacht of the Royal Yacht Squadron.

*Terra Nova* was taken in tow by Mr Edmund Handcock's tugs, the *Falcon* and the *Bantam Cock*. The farewell fleet of numerous small craft was accompanied by the paddle steamers *Devonia* and *Ravenswood*, each packed with passengers joyously cheering their farewells and waving flags. The tow was cast off as *Terra Nova* passed Penarth Head, the farewell fleet keeping company as far as the *Breaksea* Lightship off Barry.

Here, *Terra Nova* hove to while the ladies, the Lord Mayor's party, Daniel Radcliffe and Scott left the ship, taken off by pilot cutter. Three years later, Daniel Radcliffe recalled Scott's prophetic words as they left *Terra Nova* together, 'I will reach the South Pole or I will never come back again'.

Scott returned to London on expedition business and would rejoin the ship at Cape Town. *Terra Nova* continued on her passage south under the command of Lieutenant Evans (NB: two persons by the name of Evans were now aboard the ship, Lieutenant E.R.G.R. 'Teddy' Evans and Petty Officer Edgar Evans, who joined the expedition at Cardiff, hereinafter referred to as PO Evans).

The farewell fleet passing Penarth Head just prior to casting off from tugs *Falcon* and *Bantam Cock* and accompanied by the paddle steamers *Devonia* and *Ravenswood*. (Courtesy Penarth Yacht Club)

The first stop was to be Madeira. As promised, her captain, Lieutenant Evans kept her close to the Welsh coast for as long as he could so that sightseers could catch a glimpse before she turned to pass Lundy Island. Here they felt the first heave of the Atlantic as *Terra Nova* breasted the waves, her bowsprit rising and falling to each wave. It must have been a proud feeling for Evans to leave the Bristol Channel and make for the open Atlantic after such a splendid farewell. This, after all, was his first command and he must have felt thrilled with all the excitement and expectation of what was to come.

## Around the World to Lyttelton, New Zealand

### UK to Madeira
For the first twenty-four hours or so they were able to remain under sail, but the wind dropped and under steam power the ship gave all those not used to such things plenty of practice at keeping their feet and fighting seasickness, rolling, as she did, up to 28 degrees. In his diary, Dr Wilson had recorded this, noting that *Discovery* had rolled to 52 degrees and Lieutenant Evans had told him that *Morning* had rolled to 58 degrees.

For the first few days Dr Wilson spent much of his time rummaging around the cargo looking for his personal effects which had been sent ahead of him

AB James Skelton at the wheel. (Courtesy Eva Skelton Tacey)

by rail and post. When not looking for his kit, he was observing and sketching wildlife and he liked to sleep above decks on the ice-house.

Cherry-Garrard, in his classic story, *The Worst Journey in the World*, described the watch-keeping arrangements:

> Under Campbell, the 1st officer, the crew was divided into port and starboard watches and the routine of a sailing ship with auxiliary steam was followed. No work was definitely assigned to any individual on board and the custom of the ship seemed to develop with everybody turning their hands to everything when duty called.

It may have seemed that way to Cherry-Garrard, a member of the 'afterguard'. He recalled that constant calls to shorten sail, shift coal, shift cargo or to pump, paint or wash down the ship were responded to fully. Officers, scientists and seamen, no matter what their position, turned their hands to everything and Cherry-Garrard heard no complaints. When they had set out they were comparative strangers; in the close confines of the ship, they soon got to know each other.

After eight days at sea, they dropped anchor in Funchal Harbour, Madeira. The leak in the forepeak which was making 4¾in per hour had now reduced to 1in per hour. The expansion of the hull planking was most welcome.

## Madeira to South Trinidad Island

The ship stayed at Madeira for three days and at 5 a.m. on Sunday, 26 June, the crew weighed anchor after sending mail ashore. In a further description of life on board, Cherry-Garrard takes us to the cabins which surrounded the wardroom aft. The first on the left was that of Scott and Lieutenant Evans, but Scott was not on board and Wilson had taken his place. In the next cabin to them was Francis Drake, the secretary. On the starboard side of the screw were Oates, Atkinson and Levick, the two latter being doctors, and on the port side, were Campbell and Pennell, who was the navigator, then Rennick and Bowers, both watchkeepers. In the next cabin was Simpson, the meteorologist, with Nelson and Lillie, marine biologists. In the last cabin were Wright, the physicist and chemist, Gran the Norwegian ski expert, and himself, Cherry-Garrard, Wilson's helper and assistant zoologist.

As long as they had wind, the ship ran under sail alone. The biggest nuisance was the constant need to pump the ship. All hands were raised before 6 a.m. and the first job was the pumps. Many turned up for this job in their pyjamas and bare feet. The ship still needed an hour to an hour and a half of pumping each morning and by the time the officer of the watch was satisfied, they'd had quite enough of it.

On 1 July, Cherry-Garrard recorded that they saw the only ship ever sighted – *Inverclyde*, a barque out from Glasgow to Buenos Aires. Wilson quoted, 'Like a painted ship on a painted ocean, she lay with all sail set on an oily, calm day with a sea like glass.'

Lieutenant Evans wrote to Daniel Radcliffe on passage:

> This ship is no flyer but she is better than my last Antarctic ship, the 'Morning'. You would not know the 'Terra Nova' now, we have put some good work into her and she is as clean and in order as a Man 'o War.

North of the Cape Verde Islands in Lat. 22°28' N, Long. 23°5' W, they picked up north-east trade winds and during the day the colour of the sea changed from deep, clear blue water to a darkish thick green, brought about by the presence of large masses of plankton which drifts near the surface.

Pumping the bilges and coal trimming were clearly the worst jobs to be done. Wilson described the trimming of coal from the main hold where the bulk of it was stored. It was then moved into the two bunkers, one on each side of the engine-room stokehold. The bunkers held about 50 tons each and had to be filled periodically so that coal could be fed to the furnaces.

The work took three hours in the morning and the same in the afternoon. There was no room to stand up in the main hold until they had dug out a few tons and transferred it. It was hot work with the heat from the engine room and stoke-hold as well as their proximity to the tropics. It was a back-breaking business. They worked in total darkness except for a small oil lamp, the hold having to be aired and tested to prevent any collection of fire-damp which could cause an explosion. As in a mine, care has to be taken with coal on board ship. After their three-hour spells they became pretty tired and were black from head to foot but they were allowed an extra ration of fresh water to bathe in on the upper deck, although the coal dust took a day or two to clear out of the pores.

On Tuesday, 5 July, they had a fright, for they discovered a fire on board. Hooper opened the hatch of a store just outside the wardroom door, out of which billowed a great volume of smoke. He shouted that the ship was on fire. Lieutenant Evans, Campbell and Wilson were nearby, the ship's bell was rung 'fire quarters' and all hatches, windows and doors were shut down. Wilson got a fire extinguisher, went down into the hold and put the fire out quickly.

Paper had caught alight near straw packing, which luckily had not ignited. The fire had been caused by the carpenter, who had left a lamp alight and the roll of the ship had tipped it over. It was lucky they found the fire before it took hold, and the frightening experience was a lesson to all. The frightened culprit, who had left his cap nearby, was brought before the captain and the first lieutenant on the bridge and was 'logged'. But he was well thought of and as good as any man on the ship. The whole business was dealt with as a warning. The consequences of a more serious outcome were obvious to all.

After the fire, Wilson spent the afternoon on the pumps and in a long argument on socialism, politics and degeneracy of the nations. He recorded, 'Simpson is our only real socialist, anti-everything and a firm believer in the "Manchester Guardian".'

There were discussions on all general things and the subject of ships tonnage was debated and recorded by Dr Wilson in his diary on 7 July, in relation to *Terra Nova*. The SS *Terra Nova* had a registered tonnage of 745 tons; registered tonnage taking in only the space which can be given to cargo – i.e. pure carrying space apart from dwelling spaces. About 60 cubic feet means a ton of average goods.

In the navy, ships are always measured by their water displacement and this means an equivalent for the deadweight of a loaded ship; in the *Terra Nova* it is about 1,100 tons displacement. Nett tonnage is yet a third ship's measurement and includes only the capacity that is useful for non-consumable material, and by this the Terra Nova is 400 nett tonnage. Gross tonnage is yet a fourth, and this includes holds and coal bunkers and engine-room spaces and living spaces.

The ordinary weight used when speaking of the *Terra Nova* is the 400 nett tonnage. (The figure of 744 tons for *Terra Nova* is shown in the list of ships built by Alexander Stephen & Son, and as with most of the ships built by them at Dundee, is described as 'Builders Old Tonnage Gross', clearly a subject where there is plenty to discuss and argue about.)

The extent to which all expedition members worked together, no matter what their position or status, illustrates the teamwork that developed between them as they toiled to carry out the daily tasks aboard. The physical and mental effort involved in trimming coal and stoking the boilers was a great leveller to all and it caused a lot of sickness and discomfort to many.

Wilson wrote that Evans proved the strongest of all at stoking and completed a four-hour watch by himself without any help. (He doesn't say which Evans, but it is likely to have been PO Edgar Evans.) Rennick came up after an hour and nearly fainted on the upper deck but did go down again to finish his watch.

Wilson's experience of stoking of the boilers is summarised from his diary:

> There were enormously heavy tools to use, 'prickers' and 'devil rakes', about 12 feet of solid iron to reach the back of the furnaces. They had to throw coal into each of the three furnaces every ten to fifteen minutes and this in the front of a heat and glare that scorches the eyes and made them burn. They had to spread the fire in the furnace and keep each one bright over the bars which ran back 7 or 8ft and are about 3 to 4ft across the front. The fire had to be kept glowing about 6 inches deep all over this. They had to break the clinker with a 'devil', an iron rake with three teeth which they could barely lift off the ground after two or three hours in the heat. While they were shovelling in coal they got stung by what are called 'stokehold flies' which made them jump. They are drops of hot oil off the engines which caught them on the back of the neck as they stooped to shovel. They had to get the coal out of the side bunkers, breaking up the large lumps with a sledge hammer and then, dragging it across in a bucket of solid iron, it was fed to each furnace in turn. They raked out the ashes and sent them up on a hoist. Leaning on anything to rest for a while would result in a burn or a scold.

Three hours of all that, with just a mug of cocoa, was described as the hardest work Wilson had ever put himself to and not least the noise, which was continuous and excessive, so much so that they could only shout to each other.

Cherry-Garrard mentioned that they had the pleasure of being able to spot wildlife, something that can only really be appreciated from a sailing ship. After all that work, being off watch and enjoying the wildlife must have felt like heaven. They saw whales, dolphins, Portuguese men o' war, flying fish, Mother Carey's chickens (storm petrels), shearwaters and some gannets, the latter not normally deep sea birds, which made the observers wonder where the gannets had come from.

After 7 July, they lost the north-east trade winds but had no cause to complain about the weather. They crossed the equator on 15 July and the usual rituals were enacted with much hilarity and the day ended with a sing-song in the fo'c'sle as they witnessed a magnificent sunset.

On Sundays, church services were always held and it was also the day for 'rounds', i.e. inspection of the ship, all clearing up having been done the day before. The weekly inspection consisted of the captain, first lieutenant, senior medicals, officers and warrant officers all taking part. The tour was of the men's quarters, the engine room and all around the ship.

They alternated between engine and sail according to the conditions, and when under combined power of both, they could make 9 knots. In their best run, they made 191 nautical miles in twenty-four hours. They passed through a long line of orange yellow sea, the colour altering from deep blue. It was caused by fish ova – a small sausage-shaped mass of jelly crammed full of orange yellow dots and otherwise transparent. Millions of eggs coloured the sea, floating near the surface.

Time was spent preparing the 20,000 fathoms of wire for the deep-sea sounding machine.

On 25 July they arrived off South Trinidad Island, near the Martin Vaz Islands, 680 miles east of Brazil. On the south-west side they anchored 700 yards offshore. Cherry-Garrard described the island as difficult to access with a steep, rocky coast rising to a height of 2,000ft with a big Atlantic swell.

There was much to explore from the scientific point of view and equipment was made ready. A boat was launched and a party set off to find a suitable landing place. A suitable point was found, and the main party was landed.

When they returned to the ship, a bigger swell was running onshore. They had left it late, which created difficulties in getting aboard the boats and returning to the ship. The boats were moored offshore and, in their struggles to get through the surf, the collected samples and some equipment were lost. Two men had to be left on shore for the night and after much difficulty and further loss of items, they were eventually able to return to the ship.

After two nights at anchor off South Trinidad Island, the crew weighed anchor on 28 July and prepared for the passage towards Cape Town.

## South Trinidad Island to Cape Town

The passage to Cape Town was very much different from what they had already experienced, the conditions being described by Lieutenant Evans as harder seas and heavy rolls. There were now forty men on board. Although the expedition eventually consisted of a total of sixty-five men who went south, this number was made up to include a number of changes at New Zealand, mainly crew members and those officers and scientists who joined the ship by arrangement at Cape Town, Melbourne and in New Zealand.

As they used the coal and patent fuel the ship leaked less, but there was still the pumping to do which kept the men fit. Campbell, the first officer and known as the 'Wicked Mate', was respected and admired by all as a fine seaman, though Cherry-Garrard wrote that he was always frightened of him. The scientists and non-sea-trained staff, known as the 'afterguard', willingly pocketed their pride on deck and were hustled around with the youngest seamen on board, but when below in the wardroom they argued their corner in every way.

Oates, Atkinson and Gran, 'the three midshipmen', were confirmed in their rank and a ship's biscuit broken on the head of each, in accordance with gun-room practice. After this day, during good and bad weather, these three kept regular watch with the seamen, going aloft, steering and taking all the usual duties in their turn.

At one time or another, all the men found themselves stoking side by side with the firemen and, in this fashion, officers, seamen and scientific staff formed a greater friendship and respect for one another. The long sea voyage knitted the men together and brought out hidden qualities and defects, '... not that there were many of the latter', wrote Lieutenant Evans.

Wilson was to express his view as to which of the wardroom atmospheres he preferred, that of *Discovery* or *Terra Nova*. He said that he preferred *Terra Nova*, as while there was plenty of verbal teasing and ragging, at other times there remained a certain formality, though sometimes an enforced courtesy, which led to better discipline. He felt there were none of the tensions and undercurrents of feeling that there had been aboard *Discovery*.

Wilson always found himself getting up early, finishing off more grouse reports. He was desperate to get them completed. There were sudden heavy squalls and the conditions changed from the sultry heat of the doldrums to delightful storms of wind and drenching rain, which was eagerly caught in buckets and containers for washing. They were always short of fresh water, which was being rationed to half a gallon a day each since leaving Madeira, where they had refused to take on water because of a typhoid outbreak some time before their arrival.

Cherry-Garrard's description of 'making our easting down' is superb and in trying to compress it I cannot fully do his story justice. The author feels that he would be wasting good reading material for the romantic armchair sailor who might like to get into the story himself:

A Squall strikes the ship! Two blasts of the whistle fetches the watch out, and orders are given … '*Stand by topsail halyards*', … '*In inner jib* …' sends one hand to one halyard, the midshipman of the watch to the other and the rest onto the fo'c'sle and to the jib downhaul. Down comes the jib and the man standing by the fore topsail halyard, which is on the weather side of the galley is drenched by the crests of two big seas which come over the rail.

But he has little time to worry about things like this, for the wind is increasing and … '*Let go topsail halyards*' comes through the megaphone from the bridge and he wants all his wits to let go the halyard from the belaying-pins and jump clear of the rope tearing though the block as the topsail yard comes sliding down the mast.

'*Clew up* …' is the next order and then … '*All hands furl fore and main upper topsails…*' and up we go out on to the yard. Luckily the dawn is just turning the sea grey and the ratlines begin to show up in relief. It is far harder for the first and middle watches, who have to go aloft in complete darkness. Once on the yard you are flattened against it by the wind. The order to take in sail always fetches Pennell out of his chart-house to come and take a hand.

The two sodden sails safely furled, luckily they are small ones, the men reach the deck to find that the wind has shifted a little farther aft and they are to brace round. This finished, it is broad daylight and the men set to work and coil up preparatory to washing decks!

It is eight bells, and the two stewards are hurrying along the decks, hoping to get the breakfast safely from galley to wardroom. The hourly and four-hourly ship's log is being made up force of the wind, state of the sea, height of the barometer and all the details which a log has to carry including a reading of the distance run as shown by the patent log line.

Chief Engine Room Artificer William Williams. (Herbert Ponting)

Suddenly, there is a yell from somewhere amidships ... 'STEADY' ... A stranger might have thought there was something wrong, but it is a familiar sound, answered by a 'STEADY IT IS, SIR' from the man at the wheel and an anything but respectful, 'One – two – three STEADY'... from everybody having breakfast. It is Pennell who has caused this uproar and the origin is as follows ... Pennell is the navigator and the Standard Compass owing to its remoteness from iron in this position, is placed on top of the ice-house. The steersman, however, steers by a binnacle compass placed aft in front of his wheel. But these two compasses for various reasons do not read alike at a given moment, while the Standard is the truer of the two.

At intervals then, Pennell or the officer of the watch orders the steersman to '*Stand by for a steady*' and goes up to the Standard Compass and watches the needle. Suppose the course laid down is S.40 E. A liner would steer

AB James Skelton with Bosun Alfred Cheetham. (Courtesy Eva Skelton Tacey)

almost true to this course unless there was a big wind or sea. But not so the old *Terra Nova* Even with a good steersman the needle swings a good many degrees either side of the S.40 E. But as it steadies momentarily on the exact course, Pennell shouts his '*Steady*', the steersman reads just where the needle is pointing on the compass card before him, say S.47 E and knows that this is the course which is to be steered by the binnacle compass.

Pennell's yells were so frequent and ear-piercing that he became famous for them and many times in working on the ropes in rough seas and big winds, we have been cheered by this unmusical noise over our heads.

Wilson wrote:

Pennell, who is always up before sunrise to take stars for purposes of navigation ... Pennell is at work from 4 am as a rule until past 10 pm and I have never yet known him sleep in the afternoons. He gets though perfectly extra-ordinary amount of work every day always cheerful and genial and busy, but never too busy to talk birds and to be interested in other work than his own.

Lieutenant Evans, conscious of the ship's passage having been slower than expected, wrote to his friend Percy Lewis at Cardiff, 'At sea, 13th August, 1910. We are 500 miles out from Simonstown and we shall be a fortnight overdue. We counted on her steaming 7 knots easily, but she only does 6 and she won't sail well.'

Rennick wrote to Daniel Radcliffe:

> We are nearing Capetown [*sic*] after a spell of 16 days overdue, we can't get any speed out of the old bug trap. It is only by hanging on to and risking canvas that we can manage 9 knots out of her under sail' and went on... owing to being delayed in our programme we are cutting out Melbourne and Sydney and going from the Cape to Lyttelton.

On Monday, 15 August they arrived at Simons Bay. Wilson recorded that there had been a very heavy gale again all night from the north-west, with high seas and exceedingly heavy squalls. They had to take in all sail except foresail and lower topsails. To prepare for their arrival at Cape Town, they worked hard all day in the wardroom, scrubbing and painting, completing the work just in time as they arrived at Simons Bay and within signalling distance from two naval ships, *Pandora* and *Mutine*. Wilson was able to spot his wife and Lieutenant Evans' wife aboard *Pandora*. Captain Scott rejoined *Terra Nova*, having travelled to Cape Town aboard the Royal Mail Steamer *Saxon*.

The decision as to whether or not to go to Australia was all to do with keeping up time, but Scott was determined they should go to raise desperately needed funds, despite the fact that he had already been told by the government there would be nothing for them. Scott, however, was fully determined to put in at Melbourne. He was desperately concerned over finance.

Dr Wilson left *Terra Nova* and with his wife, Mrs Scott and Mrs Evans, made his way to Australia by the Royal Mail Steamer *Corinthic* to search out and make contacts for funding. Scott was now aboard *Terra Nova*, which gave him the first opportunity on the long passage to Australia to get to know expedition members at close hand.

## Cape Town to Melbourne

Many repairs and fittings had been carried out aboard by men from the two naval ships in the harbour who had sent fatigue parties, thus allowing the *Terra Nova*'s crew a certain amount of freedom ashore, which was much appreciated. The ship left Simons Bay on 2 September 1910. Lieutenant

A relaxing scene on deck as *Terra Nova* sailed south. (Hon. Edward Broke Evans)

Evans described the passage from Cape Town to Melbourne as 'disappointing' on account of the absence of fair winds. They had a few gales but finer weather than expected.

However, the steadiness of the ship did give them the opportunity to work on plans for depots and sledging journeys. It also allowed them to complete much scientific work. Cherry-Garrard describes the mixture of weather as they continued their 'easting down' and the huge swells behind them, an awesome sight in a small ship as they rose on the crest of one great hill of water with another building a mile away:

> At times these seas are rounded in giant slopes as smooth as glass; and others curl over, leaving a milk-white foam and their slopes are marbled with a beautiful spumy tracery. Very wonderful are these mottled waves; with a following sea, at one moment it seems impossible that the great mountain which is overtaking the ship will not overwhelm her, at another it appears inevitable that the ship will fall into the space over which she seems to be suspended and crash into the gulf which lies below.
>
> The seas are long with the *Terra Nova* rolling constantly to 50 degrees and sometimes to 55 degrees. The cooks had a bad time in the little galley on the deck. Archer's efforts to make bread sometimes ended in the scuppers and

the occasional jangle of the ship's bell gave rise to the saying, 'a moderate roll rings the bell and a big roll brings out the cook'.

On 18 September they were in Lat. 39° 20' S, Long. 66° 9' E after a good twenty-four-hour run of 200 miles. They had hoped to make a landing on St Paul, an 860ft uninhabited island, 2 miles long by 1 mile wide formed by an old volcano, but a gale came up and dashed all hopes of making a landing.

On two consecutive days, they made runs of 119 and 141 miles, but on the third made only 66 miles, at times becoming becalmed so that steam had to be raised. The quieter conditions were fine for observing wildlife and working on scientific projects, but all, not least Scott, were becoming anxious to push on.

Cherry-Garrard tells of events a few days before their arrival at Melbourne:

We were doing 7.8 knots under sail alone, which was very good for the old 'Terra Push', as she was familiarly called and we were just 1000 miles from Melbourne.

By Saturday night we were standing by top-gallant halyards. Campbell took over the watch at 4 am on Sunday morning. It was blowing hard and squally, but the ship still carried topgallants. There was a big following sea.

At 6.30 am there occurred one of those incidents of sea life which are interesting though not important. Quite suddenly the first really big squall we had experienced on the voyage struck us. Topgallant halyards were let go, and the foretopgallant yard came down, but the main topgallant yard jammed when only half down. It transpired afterwards that a gasket which had been blown over the yard, had fouled the block of the sheet of the main upper topsail. The topgallant yard was all tilted to starboard and swayed from side to side, the sail seemed as though it might blow itself out at any moment and was making a noise like big guns and the mast was shaking badly.

It was expected that the topgallant mast would go, but nothing could be done while the full fury of the wind lasted. Campbell paced quietly up and down the bridge with a smile on his face. The watch was grouped round the ratlines ready to go aloft and Tom Crean volunteered to go up alone to try and free the yard, but permission was refused. It was touch and go with the mast and there was nothing to be done.

The squall passed, the sail was freed and furled and the next big squall found us ready to lower upper topsails and all was well. Finally the damage was a split sail and a strained mast.

The next morning a new topgallant sail was bent on, but quite the biggest hailstorm I had ever seen came on in the middle of the operation. Much of the hail must have been inches in cumference [sic] and hurt even through thick clothes and oil-skins. At the same time there were several waterspouts formed. The men on the topgallant yard had a beastly time. Below on deck, men made hailballs and pretended they were snow.

From then onwards, they ran the course to Melbourne before a gale and on the morning of 12 October the Cape Otway Light was in sight. That evening, as it blew hard, the *Terra Nova* made Melbourne Harbour.

The officer engineer appointed for the ship was Lieutenant Edgar Riley, but he was described by Charles 'Silas' Wright in his diary as 'always sad, has supreme contempt for sails'. Simpson's journal records, 'Riley, our engineer has not been equal to the strain, he has had bad health all the way out.' On arrival in Australia, he was invalided out of the expedition. Responsibility as chief engineer in the engine room was then taken by Chief Artificer William Williams.

Since arriving in Melbourne, having travelled down in RMS *Corinthic* with his wife, Oriana and Mrs Scott, Dr Wilson had been busy meeting many Ministry and private secretaries, state governors and their aides in an effort to raise finance. He had met the prime minister (federal), Rt Hon. Andrew Fisher, the premier who had written six weeks previously to Scott to tell him to expect nothing from Australia. Before the expedition departed Melbourne, after much hard talking by Scott and Wilson, as it turned out they got all they asked for, namely £5,000 in all, £2,500 from the federal government and £2,500 from a private individual, Samuel Hordern, head of a large Sydney department store.

Wilson wrote:

> It was worth going to Melbourne for, the thing was really worked for us by Professor David* and Professor Spencer, and Lieutenant Evans as a Welshman with his arm around the Minister's waist his name being Hughes and his nationality in accordance with his name! This got us the first half and the other half was given by a man who was disgusted with his Government for not giving us the whole sum at once!

At Melbourne waiting for Scott was that dramatic telegram: 'Madeira. Am going South. AMUNDSEN.'

On the evening of 17 October, *Terra Nova* anchored close to the warship HMS *Powerful* (13,000 tons) and the admiral and his officers came aboard and

---

\* Professor Edgeworth David FRS, member of Shackleton's Antarctic Expedition 1907–09, aboard *Nimrod*.

The ship's cat in his bunk. (Courtesy Scott Polar Research Institute)

were taken around. Many searching questions were asked and even the ship's cat was inspected. Wilson's diary goes on:

> We have a small muscular black cat called 'N*****' who came on board as an almost invisible kitten in London or Cardiff! He has grown stiff and small and very strong and has a hammock of his own with the 'hands' under the fo'c'sle. The hammock is about 2ft long with proper lashings and everything made of canvas. A real man o' war hammock with small blankets and a small pillow and the cat was asleep in his hammock with his black head on the pillow and the blankets over him. The Admiral was much amused and while he was inspecting the sleeping cat, ... opened his eyes, looked at the Admiral, yawned in his face and stretched out one black paw and then turned over and went to sleep again. It was a very funny show and amused the Admiral and his officers as much as anything. [The cat] has learned to jump into his hammock which is slung under the roof with the others and creeps in under the blankets with his head on the tiny pillow.

More of the ship's cat and his adventures later in the story.

## Melbourne to Lyttelton

They were now ready to leave Melbourne and when the admiral's party left the ship they got into their old rig and prepared to make sail for New Zealand. Here, at Melbourne, Captain Scott, with the ladies, left the ship and went on to New Zealand by passenger liner, the SS *Warrimoo*, so that they could resume the begging campaign. For the passage to Lyttelton, Wilson rejoined the ship.

It was a coincidence that Lieutenant Evans, in command of the *Terra Nova*, had been told many years ago when he was a midshipman by that very man, now an admiral, that he had better leave the service as he would never do well in it! Lieutenant Evans now took the opportunity, with a fair wind, to show just how he could manage a ship under full sail. They sailed down the line of battleships, vigorously cheered by each, and when *Terra Nova* passed the flagship the rigging was manned, and they gave a huge cheer to the admiral's ship.

They made a splendid start for New Zealand, using wind and steam power to navigate the notorious Bass Straights where there are many islands and an area known for the frequency of fog and bad weather. While on passage, Wilson recorded that the Norwegian, Tryggve Gran, had complained about making cocoa for the morning watch as he felt that he was being treated by Campbell as a 'drumstick', which he didn't like. They realised that what Gran meant was a 'domestic', and so Cherry-Garrard volunteered to become the 'drumstick' and made the cocoa. 'We are a very merry party,' recorded Wilson.

They had to sail around the southern coast of New Zealand and northwards up the eastern coast before arriving at Lyttelton, where they knew they would be welcomed as had been the other expeditions before them. They sailed through Lyttelton Heads on 28 October and the plan was that they would depart for the Antarctic on 27 November. On the evening of the 28 October Scott left Wellington for Lyttelton, travelling on the ferry, SS *Maori*.

The last four weeks at Lyttelton included strenuous work unloading the ship and systematically reloading her for the final leg of their journey to Antarctica. Of great concern was the still-leaking hull and the servicing of the pumps, so the ship was dry docked.

To the task of sorting the leak came Mr H.J. Miller of Lyttelton, who had worked on previous expedition ships. Scott wrote:

> Our good friend, Miller, attacked the leak and traced it to the stern. We found the false stern split and in one case a hole bored for a long stern through-bolt which was much too large for the bolt ... The ship still leaks but the water can now be kept under with the hand pump by two daily efforts of a quarter of an hour to twenty minutes.

Captain Scott with officers and scientists (left to right): Lieutenant Pennell, Dr Wilson, Lieutenant Campbell, Captain Scott, Charles Wright, Lieutenant Evans, A. Cherry-Garrard, George Simpson, Edward Nelson, Lieutenant Bowers (seated). (Courtesy *Western Mail*)

Coal stocks were topped up, giving the ship a total of 425 tons in her holds and bunkers. Thirty tons of coal were stored on deck in sacks, and there were 128 cases of petrol, oil and paraffin in wooden cases. One hundred and fifty frozen sheep and three bullocks were packed into the reinsulated ice house. Stalls were built under the forecastle for the nineteen ponies and thirty-three dogs brought by Meares and Bruce, who now joined the expedition, together with Anton and Demetri, dog handlers brought from Russia. The dogs were chained on top of the sacks and cases and on top of the ice house, all far enough apart to try and stop them from attacking each other. Also on deck were three motor sledges in wooden crates covered with canvas. All were roped and chained down in every possible way.

The seamen were employed in preparing sledging gear, PO Evans prepared sewing outfits, the cooks assembled mess traps and cooking utensils and Levick, Atkinson and Wilson prepared the medical kits and small surgical outfits for the sledge parties. Bowers, the stores officer, sorted and relisted everything and had the remarkable ability of remembering where everything had been placed. At Lyttelton, the expedition was also joined by more members of

*Terra Nova* in Lyttelton Harbour, New Zealand. (Canterbury Museum)

the 'afterguard', the geologists, Raymond Priestley, Griffith Taylor and Frank Debenham, together with Bernard Day, the motor engineer, and Herbert Ponting, the photographer.

The huts were partly erected on the quayside and all the beams and joints marked. Tongue-and-groove boards were cut to measure. All this was supervised by Frank Davies, the ship's carpenter, assisted by Robert Forde, George Abbott and Patrick Keohane.

Mr Joseph J. Kinsey acted as the expedition's agent in New Zealand, as he had done for previous expeditions. Scott was greatly indebted to him, and the Scotts stayed at Kinsey's House, 'Te Han' at Clifton where they could have necessary consultations on expedition matters. Scott wrote in his journal:

> His interest in the Expedition is wonderful and such interest on the part of a thoroughly shrewd businessman is an asset of which I have taken full advantage. Kinsey will act as my agent in Christchurch during my absence; I have given him an ordinary power of attorney and I think have left him in possession of all the facts. His kindness to us was beyond words.

Joseph Kinsey (1852–1936) was head of Kinsey & Co., Shipping Agents, of Christchurch and the Belgian Consul for New Zealand. He was born in Kent and was for some years a master at Dulwich College. He was knighted in 1918 for his services to British Polar exploration in the Antarctic.

The ports of Lyttelton and Port Chalmers of South Island, New Zealand, became part of the great story of the heroic age of Antarctic exploration. These were the natural last ports of call before departure to the Southern Ocean and the Ross Sea region, the area to be explored. The people of New Zealand, with their scientific facilities, warmth, enthusiasm, willingness and generosity, desired to be part of this great enterprise. This, coupled with the professional assistance given by harbour board and generous local companies, gave confidence and morale-boosting support to both the polar expeditions led by Scott and the expedition led by Shackleton. At that time, extending the boundaries of the earth engaged the interest of the whole western world.

## Lyttelton to Port Chalmers and the Last Farewell.

On 25 November 1910, with the harbour steam tug *Lyttelton* in attendance, *Terra Nova* departed Lyttelton to make a call further down the coast, where they were to take on the last of the coal stocks at Port Chalmers. When they departed, the ship was given a tremendous farewell with bands playing on the harbour accompanied by rapturous cheering. Ships flew farewell messages, which Lieutenant Evans said they did their best to answer.

Herbert Ponting, the camera artist (a title he preferred), had now joined the expedition and he was busy aboard with his cinematograph. Scott and his wife went out as far as the Heads and returned in the harbour tug, *Lyttelton*. They then both journeyed to Port Chalmers by train, 100 miles down the coast to join the ship.

At Port Chalmers, Scott recorded that he met George Fenwick, managing director of the *Otago Daily Times*, to discuss the Central News Agreement, and went to the city to thank Mr Robert Glendenning for the handsome gift of 130 grey jerseys. Scott also met the mayor and drank toasts to the success of the expedition. Many other gifts of food and equipment were received, not least the generous quantity of coal put aboard by John Mill & Co. At Dunedin, the Scotts were the guests of Mr & Mrs William Moore at 'Venard', their Mornington home.

Herbert Ponting writes poetically in *The Great White South* as he describes the SS *Terra Nova* awaiting departure:

She was a picturesque sight as she lay alongside the quay at Port Chalmers. Confidence in her staunchness and ability for the tremendous task that lay ahead was bred in the knowledge of her years of fighting with the polar ice, recorded in the log-books that formed the history of her gallant and honourable past; and imagination conjured up many a brave and thrilling fight with tempestuous seas of which her figure-head might tell, could those parted lips but speak.

Herbert Ponting's book is beautifully written and a joy to read.

# 6

# INTO THE SOUTHERN OCEAN

## Final Preparation and Departure from Port Chalmers, New Zealand

On Tuesday, 29 November 1910, the expedition departed Port Chalmers at 2.30 p.m. in bright sunshine. The ladies and Mr Kinsey, travelled as far as the Heads and returned in the harbour tug, *Plucky*. At both Lyttelton and Port Chalmers the expedition members were greatly touched by the warmth and overwhelming generosity of the New Zealand people. Scott had previously experienced the kindness and help given at Lyttelton and Port Chalmers when on the *Discovery* expedition. Polar explorers of the heroic age and those in later times shared this great tradition, which had begun in the nineteenth century. The bays and harbours of New Zealand were for the 'last farewells', and particularly so Port Chalmers, as it is the last port of call before the final passage into the Southern Ocean. The story of Port Chalmers and the connection with many Antarctic expeditions is told in *Last Port to Antarctica* by Ian Church.

With the cheers of the people behind her, *Terra Nova* left Port Chalmers and Lieutenant Evans recorded that he had a 'heart like lead'. He felt an awful feeling of loneliness, no doubt shared by many. But the crew set about their work to hide their feelings as New Zealand faded away, with the spring night gently obscuring the land from view.

Tryggve Gran recorded in his diary:

The final goodbyes at Port Chalmers Pier, 29 November 1910. (Courtesy Port Chalmers Museum)

All links with civilization are cut, and as night falls New Zealand sinks from sight. It is almost sad to think that years will pass before we shall once more see land with forests and green fields. We left Port Chalmers about 2 pm and sailed out into the fjord to the cheers of thousands of onlookers. We now lie off the New Zealand coast and are adjusting compass. (Midnight) The Dunedin light is on the horizon. It is a wonderful night, starlit and mild. About 9 pm a breeze sprang up from land, and we are southward bound under full sail towards the unknown.

The following day, Gran made a more cautionary entry in his diary, 'It was beautiful weather this morning, but this evening there is a fresh breeze from the west and the sky is dark and threatening. The ship is overloaded, so let's hope we don't run into bad weather.' He added, '(Had it not been for the Royal Yacht Squadron ensign waving at the spanker gaff, the craft would never have received permission to sail out to sea – TG)'.

Evans and Scott thought highly of the *Terra Nova*'s bosun, Alfred Cheetham, and the seamen crew. Lieutenant Evans recorded that they could not make out how 'Alf', as the sailors called him, got so much out of the hands:

The last farewells at Port Chalmers, New Zealand. (Courtesy Port Chalmers Museum)

This little squeaky-voiced man … I think we hit on Utopian conditions for working the ship. There were no wasters and our seamen were the pick of the British Navy and the Mercantile Marine. Most of the Naval men were intelligent petty officers and were as fully alive as the merchantmen to 'Alf's' windjammer knowledge.

Cheetham was quite a character and, besides being immensely popular and loyal, he was a tough, humorous little soul who had made more voyages below the Antarctic Circle than any man on board and thereafter in the heroic age.

Cheetham had been trained before the mast and was in the *Morning* to relieve *Discovery* and had been in *Nimrod* with Shackleton. After the *Terra Nova* 1910 expedition, he went with Shackleton again, on the *Endurance*, and was one of the men rescued from Elephant Island. Sadly, he was lost in the English Channel, his naval vessel destroyed by enemy action toward the end of the First World War.

Herbert Ponting, in *The Great White South*, tells us of his work as camera artist. His laboratory was situated in a deckhouse, on the port side forward of the poop. He describes the difficulties of coping with a photographic studio at sea, which he himself had planned and supervised in its construction. Every time the seas got up streams of water squirted through chinks in the badly fitted door and showered him through the mushroom ventilator above. He had his work cut out to save his photographic gear from ruin.

Despite all this, he found enough comfort there to make room for his bunk on the floor rather than sleep in the congested cabin space he had been allocated. Ponting was neither a seaman nor a scientist in expedition terms, but he had travelled a number of times around the world visiting many foreign places. His independent duties as photographer enabled him to make interesting observations and assessments of the expedition members living and working closely together, both aboard ship and ashore.

He was already established as a photographer, and his Antarctic photographs are now famous. He wished to be known as a 'camera artist' and was nicknamed 'Ponko'. Much of his story tells of the wildlife around the ship and he describes the cape pigeons, mollymawks, snowy white petrels, pintado petrels, Wilson storm petrel\* or 'Mother Carey's chickens', as they used to be known,

---

\*   Named by Alexander Wilson, the early nineteenth-century Scottish American ornithologist.

*Terra Nova* departing from New Zealand, 29 November 1910. (Courtesy Hon. Edward Broke Evans)

Captain Scott with officers and scientists in the wardroom. (Herbert Ponting)

circling *Terra Nova*. Not least, the most majestic birds of all, the magnificent albatrosses more than 10ft from tip to tip, which soared like aeroplanes about the wake of the ship, never flapping their wings but simply setting them against the breeze. In his enthusiasm to take photographs, he nearly lost his finest hand camera overboard, falling amongst timber stowed on board and in doing so nearly breaking his leg.

## A Storm in the 'Furious Fifties'

As they progressed south, the overladen ship was a cause of anxiety to all and the last thing they wanted was bad weather, but they got it. On Thursday, 1 December 1910, they ran into a gale which, in Ponting's words, was blowing 'great guns from the west'.

Good progress had been made through the 'roaring forties', now they were in the 'furious fifties'. They shortened sail to lower topsails, jib and stay-sail in the afternoon but both wind and sea rose with great rapidity. Despite all the careful lashings the deck cargo began to work loose. They were forced to throw sacks of coal overboard which were acting like battering rams, and they relashed what they could. Priestley was the most seriously incapacitated by seasickness. Ponting could not face meals but struggled on.

Tryggve Gran recorded in his diary for 1 December:

I am wet to the skin after standing watch. The weather has been foul, with high seas and decks awash for long periods. The wind blew up in the morning watch and the sea got up unbelievably fast. At about 5 pm we had a hell of a job with the sails; all but the staysail had to come down. With the shortening of sail, 'Terra Nova' lay almost dead in the wave troughs and great masses of water washed over the decks. The deck cargo had to be relashed and we worked for hours almost like divers. The night watch has been no better. The coal sacks came adrift and had to be thrown overboard. And what a struggle it was to dump such a valuable cargo! This heaving and hauling needed great care. We had to wait for the ship to steady after each wave before getting a grip on the sacks. Last night, however, the waves often surprised us and we had to be really quick. Between coal sacks deep under water, we were hurled from rail to rail, but now it is over for a while. The bunk awaits. That will be good.

Seamen crew of *Terra Nova*, 1910 (left to right). Back row: Horton, Brissenden, Parsons, Heald, Neale, Balson, McCarthy. Centre row: A. McDonald, W. McDonald, Burton, McGillon, McKenzie, Omenchelco, Clissold, Mather, Davies. Front row: Skelton, McLeod, Bailey, Forde, Leese. (Herbert Ponting)

As the night wore on, they were forced to hove to. The bad news from the engine room was that the pumps had choked and water had risen over the gratings. Lashly and Williams were up to their necks clearing the pump suctions, but the handpump produced only a dribble. Water was creeping up to the boilers and Williams, the chief engineer, had to admit that he was beaten. The main bilge pump was dependent on the engine and to work it they had to keep up steam. The hand pump which operated from the deck was also blocked and produced only a trickle. If the water reached the boilers, then the casings would crack and might explode. Permission was given to Williams for fires in the boilers to be drawn. This meant that engine power was lost.

The seas were sweeping over the deck from the fore rigging to the poop deck and the lee rail was well under. The afterguard was organised into two parties by Lieutenant Evans and all they could do was bail out the engine room with buckets, while work continued below to try and unblock the pumps. The overloaded ship was being strained to its limits, the decks were leaking badly, and the ship was filling up. Griffith Taylor wrote of 'something unique in modern times, in a ship of 750 tons ... bailing with buckets!'

On 2 December, Gran recorded in his diary:

What a state we're in! To keep the vessel afloat we have bailed with buckets since four this morning. The pumps have stopped, for the water has long since doused the boiler fires. I went down to the engine room this morning and to my astonishment found the crew down there working in water up their necks.

Throughout the mayhem, Oates and Atkinson cared for the ponies, while Meares and others were trying to prevent the dogs being hanged by their chains. One dog was washed overboard and the next wave washed him back on again. The sight of the animals in their plight upset Scott, who could see the distraught look in their eyes. The stewards, Hooper and Neale, despite all the difficulties, carried on superbly with their duties in the wardroom and the cooks, Archer and Clissold, kept the galley fires going and provided hot cocoa.

Access to the hold to clear the pumps would normally have been through the aft hatch up on the deck, but opening it was impossible because of the amount of sea running over the ship. Had that hatch been opened, the ship would have gone down in minutes. Instead, they got to the base of the pump wells from below by cutting a hole through the engine room bulkhead, giving access to the hold and the pump well.

Cutting through the iron bulkhead was begun by Chief Engineer Williams and the ship's carpenter, Davies. It took twelve hours to cut the hole and gain access, while the men bailed the engine room with buckets. Eventually the pump wells, thick with a mixture of coal dust and water, were unblocked. The pumps were brought into action, the level of the water below was lowered and the engine restarted. It was a great relief to all when the weather subsided.

Scott recorded that he was proud to be wonderfully served by such men, who sang sea shanties throughout the difficulties. Losses were listed as two ponies, one dog, 10 tons of coal, 65 gallons of petrol and a case of biologists' spirit. They were fortunate not to have lost a man overboard in those conditions.

This particular event in the story of the *Terra Nova* goes far deeper than just a storm at sea for, in looking back, the whole expedition might have been lost without trace. Putting an expedition together virtually 'on a shoestring' resulted in overloading a ship beyond maritime regulations to avoid that which, for years before, had been fought for and won for seamen on safety grounds. To sail, as on this occasion, under the White Ensign and the burgee of the Royal Yacht Squadron may be described as true 'flag of convenience' sailing. It was taking a hair-raising chance with men's lives in known treacherous seas, all in pursuit of ambition, adventure and conquest. This was pushing seamanship and exploration to its limits.

Manning the pumps in a storm. (Herbert Ponting)

In this particular adventure the whole expedition did survive, and it will always remain a lesser-told part of this story of the heroic age. We find risky exploits in other expeditions risky too, in quests to reach both the north and south extremities of the globe. We take for granted the technology of the later twentieth and the early twenty-first century with its built-in global navigation systems enabling communication and rescue capability. No such help existed for these men. They were pushing the limits of what was available in their day. There can only be admiration for men who fought to survive in those conditions.

More detailed accounts of that storm between 1 and 3 December 1910 are recorded by Wilson in his diary, in a letter written home by Bowers to his mother and by Scott in his journal. In their accounts emerge a record of a desperate fight against the elements which they overcame by leadership, teamwork and sheer tenacity. After all, there was nowhere else to go except to the bottom of the Southern Ocean. They had to fight to keep their ship and to survive.

Wilson's diary tells of those thirty-six hours and Bowers' letter contains more detail of the situation and of his part in the emergency work down below in the engine room. The diary entries by Dr Wilson for 1–3 December are reproduced below, followed by an extract from the letter of 'Birdie' Bowers written to his mother:

Thurs 1 Dec
We were to have paid Campbell Islands a visit on our way south and to have left the instruments there for a meteorological station and to have instructed some of the 12 shepherds, who live there to take observation. But we have made such a good pace south that by tonight we shall be there and as the wind is freshening we have decided not to stand off there till the morning but to leave them and go right on and establish the station there when the ship returns with Pennell. This decision was come to in the morning by the evening the weather had so far altered for the worse as to leave us no choice and we went right on under steam and full sail. So we had no sight of the Campbell Islands. I was rather sorry for this as I had hoped to see albatrosses nesting there, but the barometer is very threatening and we have no choice but to go on south as we could make no headway against the wind now to get into Campbell Islands under steam. There are no lights and the harbour is impossible at night. At 9.00 a.m. we sent off two carrier pigeons which had been brought by Paton from New Zealand. I believe he has two more which will be liberated later – no, I find they were liberated yesterday at

noon. I watched the two go off today and they looked strong enough on the wing. They flew round and round the ship in increasingly wider circles, rising to a good height and gradually disappeared somewhere in a sort of circular direction. It was impossible to say that they had gone in any particular direction. They ought at any rate to reach Campbell Islands one would think.

We are today at noon in 50° 44' S, 170° 38' E having made good 191 miles since noon yesterday and the barometer keeps falling, falling constantly and the wind and sea and swell rising all the forenoon and afternoon until it became force 8 on the scale of Conditions of the Sea (1) which I will give presently. By 4 p.m. in the afternoon it was blowing so hard that both topgallant sails were taken in, and between 4 p.m. and 6 p.m. the main and outer jib were also taken in, as well as the fore and main upper topsails, and at 6 p.m. the foresail also was taken in. The wind was W.N.W. force 6 in the forenoon and by midnight had gone through S.W. by S. force 7 and 8 to W.S.W. force 9. The sea during the day rose from 5 to 8 and we were taking in green seas on both sides filling the waist of the ship and tearing everything adrift that could come. The ship wallowed in the water like a thing without life and it was evident that we were in for trouble with a barometer steadily falling and continuing to fall and our heavy deck cargo. Besides all the ordinary ship's necessities on the upper deck we have added as permanent weights three deck houses on the poop, including all our laboratories, which are full of gear as they will hold, five boats and the ice-house. Then there are 162 carcasses of sheep in it and two of beef; three enormous cases containing the motor sledges chained down to the deck about 128 cases of petrol, oil and paraffin, in two and three layers, tied down to the deck; 15 tons of horse fodder in compressed bales; 33 dogs, and stables all solidly built of heavy timber, and 19 horses in them; 30 tons of coal in sacks thrown in at the last moment, wherever there was space to take one. We have 1,700 gallons of petrol on the upper deck and it was this, in tin cans packed two together in heavy wooden cases, that caused most of the trouble. They were all arranged as a sort of floor on the upper deck and we walked over them and over the coal sacks and amongst the dogs and the coils of ship's ropes. We climbed over the heavy wooden props and stays and chains which kept all the motor car cases in their places in fact there is no clear space on the whole deck on which to move the deck itself being invisible everywhere. And yet the heavy green seas that came on board over the rail, both on the weather side and, still more disastrously, on the lee side as she rolled to leeward wallowing into a trough of the enormous seas, got under the whole of these cases and gradually loosened the ropes and wedges which held them down. It was rather appalling to see the whole floor of these rising and getting out of place until at last one or two got adrift and then the whole lot started moving about and breaking one another up. We had to plunge every now and then into the drowned waist of the ship, seize one of the broken cases and haul the two tins of petrol 10 gallon tins – up on to the poop to save them from being burst. Constantly one was caught in the act and a green sea would sweep along the whole deck, rail under, and over one's shoulders while the cases beneath all floated up anyhow, jambing one's seaboots and themselves in every possible direction, and forcing one often to drop everything and cling to the pin rail (2), or the ropes attached to it, instead of going over the side with tons of water. I had never seen a ship wallow so persistently or take such tremendous seas over the rail every time she rolled to leeward as this ship. Of course we were much too heavily laden and the storm was a very heavy one. I have already said that by 6 p.m., the only sail we carried were the lower topsails and the fore topmast staysail but we were still under steam. Early in the afternoon the seas were coming in so heavily that 5 tons of the deck cargo of coal in sacks was thrown over the side to facilitate the job of securing the petrol cases again. By the evening the sea was tremendous, but whether it was really bigger than anything we saw in the *Discovery* I can't say. Certainly I never saw any ship so constantly swept from foc'sle to poop, or sometimes right on to the poop, by green seas. Then one of the horses went down and was got up on its legs again, then another, and another – and Oates and Atkinson worked like bricks helping them up. Then one refused to stand or be helped up, and then one died and all this happened in the afternoon of the forenoon when we were playing about with carrier pigeons and wondering what the barometer was falling so quickly to bring us. The one thing we didn't want was this heavy storm – but it came right enough and it came quickly, and when everything began to happen at once it looked as though it had come to save us a great deal of trouble and two years' work, for 24 hours after the storm began it looked to everyone who knew what was going on as though we must go to the bottom. The afternoon and evening had been wet and cold and fatiguing for everyone and we turned in for a bad night, as the ship's decks were so strained that she was pouring water into every cabin. Everything everywhere was wet and we were of course battened down yet all through the night one heard periodically the thud of a sea on the decks overhead and a water cart full of water pour down upon the wardroom floor and table and then subside into a trickle and a drip. Seas were breaking over us, especially aft, and we were very much lower there in the water than further forward. I must say I enjoyed it all from the beginning to end, and one bunk became untenable after another, owing to the wet, and the comments became more and more to the point as people searched out dry spots here and there to finish the night in oilskins and greatcoats on the cabin or wardroom floors, or on the wardroom seats. I thought things were becoming interesting. This night, however, I had a dry bunk and a good sleep until early dawn, when the storm was worse than ever. So far, however, it was only an unfortunate storm breaking on an over-laden ship, and save for the horses which were being thrown about in an unmerciful way in their narrow stalls, and for the dogs, which were miserably and constantly being nearly drowned by seas, things were shaking down in a fairly manageable way. The engines were working and the water in the well was kept well in hand by the steam pump, though the straining of the old ship was evidently helping her to take in water through the seams. Then dawn came and with it things began to go wrong again all at once.

Fri 2 Dec

We were now in 52° 07' S. and 172° 11' E. and since noon the previous day we had made 101 miles S. 35 E. We were under easy steam and main lower topsail, and foretop staysail, and at 4 a.m. furled fore lower topsail and inner jib. Then we were all turned out to lend a hand on deck, those of us who were fit for it. Some of the staff were like dead men with seasickness. Even so, Cherry-Garrard and Wright and Day turned out with the rest of us who were not seasick, and alternately worked and were sick. I have no seasickness on these ships myself under any conditions so I enjoyed it all, and as I have the run of the bridge and can ask as many questions as I choose I knew all that was going on. The only breakfast this morning was some coffee and biscuits and whatever could be discovered dry enough to eat in the pantry; it was impossible to bring food through the seas from the galley. The waist of the ship was a surging, swimming bath, and one went there at the risk of being drowned alive in one's clothes before getting out again, or else of being broken up amongst the ruck of petrol cases, or else of mistaking the outside of the ship's rail for the inside in a green sea and not finding one's way back again by the end of whatever rope one happened to seize as the deluge, engulfed one. It was really terrific and the alternative was to climb up by the pump on to a hayrick of forage, then on to the main hatch afloat with dogs, then up on to the ice-house, and so to drop down by the foc'sle -all the while with very little to hold on to, and some parts nothing, with the chance of being swept over any moment by a sea. It was all very exciting, but wet and very cold.

Between 5 and 6 a.m. the bunkers were leaking so much water down into the engineroom that we had to pass the coal down in sacks to the stoke-hold by hand from the upper deck. The water in the stoke-hold was gradually gaining on the steam pumps chiefly because the rolling of the ship so stirred up all the coal dust in with the bilge water that it clogged in great balls in the pump valves. This entailed a very unpleasant duty for the engine-room leading stokers. Lashly, for instance, spent hours and hours up to his neck in bilge water beneath the foot plates of the stoke-hold clearing these balls of oily coal dust away from the valves which could be reached no other way. The water gained and gained until at last it was impossible to reach the valves without getting under the bilge water entirely and that then became impossible. The moment this happened the valves became so clogged that no water could be pumped out by the main engine pumps at all and we had to fall back on a subsidiary pump which was much smaller but which worked for a very short time and then also got its valves choked, also out of reach on account of the depth of water already leaked into the ship. There was now a huge wave of water in the engine-room stoke-hold, washing backwards and forwards, or rather from side to side, with a rushing roar at every roll of the ship. In this water the stokers stood and worked at the furnaces to keep the engines going – and the noise and steam and clatter of iron and rush of water down there the dark lamplight and glare of occasionally opened furnace doors was fearsome. It was the one place in the ship where the rise of water could he watched inch by inch, and it continued to so. The wind meanwhile was blowing its very hardest from the S.W. force 10, then SSW. to 10 to 9, then to S. force 9. The sea was very high indeed. I don't think I have ever seen such a high sea and such an immense swell running at such a furious rate, and the old ship simply lay down in it and wallowed every time. The port whaler – the boat hanging high up on a level with the bridge rails – was constantly dipping into the water and was lifted in the davits and swung in under the bridge deck and there jammed. The weight of water in the waist of the ship began early today to take timbers out of the rail bulwarks, and at 8 a.m. we had one of the bulwark uprights carried away and all the planking so that there is now a big gap in the bulwarks. We lost today 60 gallons of petrol and about 20 gallons of lubricating oil carried overboard in cases by the heavy seas. We also lost a dog carried overboard with a broken chain and we lost another pony which died of exhaustion after falling in its stall. They were all desperately tired of trying to keep upright in the roll of the ship which is pitching also in the most trying manner. They seem every now and again to forget and go to sleep and then the ship catches them unawares and down they go without room or strength to recover themselves.

If the old ship is opening up everywhere, as she is on the deck above our cabins, there is nothing to wonder in her making water, for she is leaking like a sieve, and at 7.00 a.m. the wardroom officers and staff had to lend a hand to try and get some of the water out by hand as both the engine pumps were out of action. So the main hand pump handles were shipped in the waist of the ship where we were at one moment up to our knees and another up to our necks in water, but no sooner had we started to pump than we found the valves there were also blocked with caked oil and coal dust. The water still rising in the stoke-hold and rushing about the foot plates in such an enormous wave and so close sometimes to the furnace gratings just above them that it was considered no longer safe to keep the fires alight for fear of an explosion of steam so the fires were drawn and the engines stopped. This was a bad state of things as the water in the hold was rising all the time. The next thing we got at was a small hand pump for two hands which again failed as the lift in the pipe was insufficient for the depth of the stoke-hold, and indeed it was not made for such a lift or ever intended for it. And last of all at 7 a.m., as a last resource, the whole of the wardroom mess, not laid out with seasickness, was divided into two watches and started baling out the stoke-hold, that is the ship! For the stoke-hold was the only place where it could be reached with buckets! All Friday and Friday night we worked in two parties, two hours on and two hours off. It was heavy work filling and handing up huge buckets of water as fast as they could be given from one to the other, from the very bottom of the stoke-hold to the upper deck, up little metal ladders all the way. One was, of course, wet through the whole time in a sweater and trowsers and seaboots and every two hours one took these off and turned in for a rest in a great coat, to turn out again in two hours and put on the same cold sopping clothes, and so on until 4 a.m. on Saturday when we had baled out between 4 and 5 tons of water and had so lowered it that it was once more possible to light fires and try the engines and the steam pump again and to clear the valves and the inlet which was once more within reach. The fires had been put out at 11.40 a.m. and were then out for 22 hours while we baled. It was a weird night's work with the howling gale and the darkness and the immense sea running over the ship every few minutes, and no engines and no sail, and we all in the engine-room black as ink with engine-room oil and bilge water, singing chanties as we passed up slopping buckets full of bilge, each man above slopping a little over the heads of all of us below him wet through to the skin, so much so that some of the party worked altogether naked like a Chinese coolie and the rush of the wave backwards and forwards at the bottom grew hourly less in the dim light of a couple of engine room oil lamps whose light just made the darkness visible. The ship all the time rolling like a sodden, lifeless log, her lee gunwale under water every time. All this time another effort was being made at Captain Scott's suggestion – and this was to reach the suction well of the hand pump through the bulkhead which divided it in the after-hold from the engine-room. It meant hours of work on top of and at the back of the boilers, which were hardly cooled, and as the bulk-head was of iron sheeting it took hours to cut a manhole through. Still it was done, and then a quantity of coal was taken out and the suction valve reached and cleared. It all took a lot of time and at this work Evans was for hours in water and in oil and coal dust, and Williams, the engineer, worked end on for hours till the thing was done by 9 o'clock on the Saturday morning.

Sat 3 Dec

We were all at work till 4 a.m. and then were all told off to sleep till 8 a.m. At 9.30 a.m. we were all on the main hand pump, and low and behold it worked and we pumped and pumped till 12.30 when the ship was once more only as full of bilge water as she always is and the position was practically solved. We had been hove to all yesterday, just wallowing in the wet, and we had moved 23 miles to leeward S. 77° E. We were now in 52° 12' S., 172° 48' E. At 2 a.m. the fore lower topsail and the main upper topsail were set as the wind had dropped to 5 or 6 from the S.S.W., though the sea was as bad as ever and the swell immense. At 6.15 a.m. the fore upper topsail and the foresail were set. The day went in pumping out the ship and securing gear and making all the moving deck cargo fast again. There was one thrilling moment in the middle of the worst hour on Friday when we were realising that the fires must be drawn, and when every pump had failed to pump, and when the bulwarks began to go to pieces and the petrol cases were all afloat and going overboard, and the word was suddenly passed in a shout from the hands at work in the waist of the ship trying to save petrol cases that smoke was coming up through the seams in the after-hold. As this was full of coal and patent fuel and was next the engine, and as it had not been opened for the airing it required to get rid of gas on account of the flood of water on deck making it impossible to open the hatchways the possibility of a fire there was patent to everyone and it could not possibly have been dealt with in any way short of opening the hatch and flooding the ship, when she must have foundered. It was therefore a thrilling moment or two until it was discovered that the smoke was really steam arising from the bilge at the bottom having risen to the heated coal. This hatchway it was impossible to open as it had tons of water washing over it at every roll of the ship and yet it was the only way to get at the suction inlet of the main hand pumps, which were choked with oil and coal dust and which, therefore, had to remain choked with coal dust until the manhole had been cut through the iron bulkhead from the engine-room into the fore-hold where the pump well was placed. Just about the time when things looked their very worst, the sky was like ink and water was everywhere, everyone was as wet inside their oilskins as the skins were wet without, and things looked very bad indeed in every way, there came out a most perfect and brilliant rainbow for about half a minute or less, and then suddenly and completely went out. If ever there was a moment at which such a message was a comfort it was just then it seemed to remove every shadow of doubt, not only as to the present issue, but as to the final issue of the whole expedition and from that moment matters mended and everything came all right. At 12.30 we set main top gallant sail, and at 1.30 the fore topgallant sail and inner jib. In the afternoon the two dead ponies were hauled up through the foc'sle skylight and passed overboard, and by night all the moveables on the upper deck were once more made secure. Air temperature 42°–44°F. Water 45°F. And as the change in these as we near the pack will interest some of you. I will put them in from now onwards.

Bowers' letter to his mother contains more details of the dangerous situation and of his part in the emergency work below in the engine room:

In the afternoon of the beginning of the gale I helped make fast the T.G. sails, upper topsails and foresail, and was horrified on arrival on deck to find that the heavy water we continued to ship, was starting the coal bags floating in places. These, acting as battering-rams, tore adrift some of my carefully stowed petrol cases and endangered the lot. I had started to make sail fast at 3 p.m. and it was 9.30 p.m. when I had finished putting on additional lashings to everything I could. So rapidly did the sea get up that one was continually afloat and swimming about. I turned in for two hours and lay awake hearing the crash of the seas and thinking how long those cases would stand it, till my watch came at midnight as a relief. We were under two lower topsails and hove to, the engines going dead slow to assist keeping head to wind. At another time I should have been easy in my mind; now the water that came aboard was simply fearful, and the wrenching on the old ship was enough to worry any sailor called upon to fill his decks with garbage fore and aft. Still 'Risk nothing and do nothing', if funds could not supply another ship, we simply had to overload the one we had, or suffer worse things down south. The watch was eventful as the shaking up got the fine coal into the bilges, and this mixing with the oil from the engines formed balls of coal and grease which, ordinarily, went up the pumps easily; now however with the great strains, and hundreds of tons on deck, as she continually filled, the water started to come in too fast for the half-clogged pumps to cope with. An alternative was offered to me in going faster so as to shake up the big pump on the main engines, and this I did in spite of myself and in defiance of the first principles of seamanship. Of course, we shipped water more and more, and only to save a clean breach of the decks did I slow down again and let the water gain. My next card was to get the watch on the hand-pumps as well, and these were choked, too, or nearly so.

Anyhow with every pump hand and steam going, the water continued to rise in the stokehold. At 4 a.m. all hands took in the fore lower topsail, leaving us under a minimum of sail. The gale increased to storm force (force 11 out of 12) and such a sea got up as only the Southern Fifties can produce. All the afterguard turned out and the pumps were vigorously shaken up sickening work as only a dribble came out. We had to throw some coal overboard to clear the after deck round the pumps, and I set to work to rescue cases of petrol which were smashed adrift. I broke away a plank or two of the lee bulwarks to give the seas some outlet as they were right over the level of the rail, and one was constantly on the verge of floating clean over the side with the cataract force of the backwash. I had all the swimming I wanted that day. Every case I rescued was put on the weather side of the poop to help get us on a more even keel. She sagged horribly and the unfortunate ponies though under cover were so jerked about that the weaker ones could not keep their feet in their stalls, so great was the slope and strain on their forelegs. Oates and Atkinson worked among them like Trojans, but morning saw the death of one, and the loss of one dog overboard. The dogs, made fast on deck, were washed to and fro, chained by the neck, and often submerged for a considerable time. Though we did everything in our power to get them up as high as possible, the sea went everywhere. The ward-room was a swamp and so were our bunks with all our nice clothing, books, etc. However, of this we cared little, when the water had crept up to the furnaces and put the fires out, and we realized for the first time that the ship had met her match and was slowly filling.

Without a pump to suck we started the forlorn hope of buckets and began to bale her out. Had we been able to open a hatch we could have cleared the main pump well at once, but with those appalling seas literally covering her, it would have meant less than 10 minutes to float, had we uncovered a hatch.

The Chief Engineer [Williams] and carpenter [Davies], after we had all put our heads together, started cutting a hole in the engine room bulkhead, to enable us to get into the pump-well from the engine room; it was iron and, therefore, at least a 12 hours' job. Captain Scott was simply splendid, he might have been at Cowes, and to do him and Teddy Evans credit, at our worst strait none of our landsmen who were working so hard knew how serious things were. Capt. Scott said to me quietly – 'I am afraid it's a bad business for us. What do you think?' I said we were by no means dead yet, though at that moment, Oates, at peril to his life, got aft to report another horse dead, and more down. And then an awful sea swept away our lee bulwarks clean, between the fore and main riggings only our chain lashings saved the lee motor sledge then, and I was soon diving after petrol cases. Captain Scott calmly told me that they 'did not matter.' This was our great project for getting to the Pole the much advertised motors that 'did not matter'; our dogs looked finished, and horses were finishing, and I went to bale with a strenuous prayer in my heart, and 'Yip-i-addy' on my lips, and so we pulled through that day. We sang and re-sang every silly song we ever knew, and then everybody in the ship later on was put on 2-hour reliefs to bale, as it was impossible for flesh to keep heart with no food or rest. Even the fresh-water pump had gone wrong so we drank neat lime juice, or anything that came along, and sat in our saturated state awaiting our next spell. My dressing gown was my great comfort as it was not very wet, and it is a lovely warm thing.

To make a long yarn short, we found later in the day that the storm was easing a bit and that though there was a terrible lot of water in the ship, which, try as we could, we could not reduce, it certainly had ceased to rise to any great extent. We had reason to hope then that we might keep her afloat till the pump wells could be cleared. Had the storm lasted another day, God knows what our state would have been, if we had been above water at all. You cannot imagine how utterly helpless we felt in such a sea with a tiny ship, the great expedition with all its hopes thrown aside for its life. God had shown us the weakness of man's hand and it was enough for the best of us the people who had been made such a lot of lately the whole scene was one of pathos really. However, at 11 p.m. Evans and I with the carpenter were able to crawl through a tiny hole in the bulkhead, burrow over the coal to the pump-well cofferdam, where, another hole having been easily made in the wood, we got down below with Davy lamps and set to work. The water

A mug of Oxo for the helmsman. (Herbert Ponting)

was so deep that you had to continually dive to get your hand on to the suction. After 2 hours or so it was cleared for the time being and the pumps worked merrily. I went in again at 4.30 a.m. and had another lap at clearing it. Not till the afternoon of the following day, though, did we see the last of the water and the last of the great gale. During the time the pumps were working, we continued the baling till the water got below the furnaces. As soon as we could light up, we did, and got the other pumps under way, and, once the ship was empty, clearing away the suction was a simple matter. I was pleased to find that after all I had only lost about 100 gallons of the petrol and bad as things had been they might have been worse …

You will ask where all the water came from seeing our forward leak had been stopped. Thank God we did not have that to cope with as well. The water came chiefly through the deck where the tremendous strain not only of the deck load, but of the smashing seas was beyond conception. She was caught at a tremendous disadvantage and we were dependent for our lives on each plank standing its own strain. Had one gone we would all have gone, and the great anxiety was not so much the existing water as what was going to open up if the storm continued. We might have dumped the deck cargo, a difficult job at best, but were too busy baling to do anything else …

That Captain Scott's account will be moderate you may be sure. Still, take my word for it, he is one of the best, and behaved up to our best traditions at a time when his own outlook must have been the blackness of darkness …

Under its worst conditions this earth is a good place to live in.

With the storm passing over, tremendous relief was felt. All aboard were now bound to have admiration and feel a trust in the ship which had brought them through the storm. After all, she was already 25 years old, little did they realise that after her time with the expedition, *Terra Nova* would continue in service for a further thirty years.

## Through the Pack Ice and into the Ross Sea

On Saturday, 3 December 1910, the wind subsided but left them with a heavy swell. They only made 23 miles that day. The state of the bunks and everything else down below was sodden. In the wardroom that night they drank to wives and sweethearts, and those not on watch were glad of an early night.

The following day gave them more fresh wind, but fine weather followed. Gran declared that he had seen an iceberg, but it turned out to be a whale spouting! The lengthening days told them that they were rapidly changing latitude and approaching the ice.

They sighted the first iceberg in Lat. 62° on the evening of Wednesday, 7 December 1910. Cheetham's squeaky hail came down from aloft and Lieutenant Evans went up to the crow's nest. Thereafter, they passed icebergs of all sizes, from the huge tabular variety to the small weather-torn bergs. By breakfast time on 9 December they were in Lat. 65° and, as well as the many varieties of birds around the ship, the first penguins and seals appeared. Those men on the ship going south for the first time, stood on the fo'c'sle and the poop, sketching, painting and enjoying the scene which unfolded before them. Scott records in his journal that they entered the pack ice on 9 December 1910, at Lat. 65° 8' S, Long. 177° 4' W '… the birds are with us and we have seen a good many whales'.

Entering the pack ice is an awesome sight when seen for the first time. For Scott, Lieutenant Evans, Wilson, Priestley, Day, Cheetham, Lashly, Crean,

Furling sails on entering the pack ice. (Courtesy Eva Skelton Tacey)

Heald, PO Evans and Williamson it was a return visit, although perhaps no less awesome.

Scott wrote often of his concern for the animals on board, he was relieved at the steadiness of the ship after their storm-tossed passage and imagined the relief and comfort it now brought to the ponies, noting that the dogs were visibly cheered and so was everybody else on board. As they crossed the Antarctic Circle at Lat. 66° 33', Scott wrote:

> The sun just dipped below the southern horizon. The scene was incomparable. The northern sky was gloriously rosy and reflected in the calm sea between the ice, which varied from burnished copper to salmon pink; bergs and pack to the north had a pale greenish hue with deep purple shadows, the sky shaded to saffron and pale green. We gazed long at these beautiful effects.

On 10 December, they took on about 8 tons of water from hummocky ice and Rennick took soundings in 1,960 fathoms and brought up two small lumps of volcanic lava. They secured four crab-eater seals and had the livers for dinner.

*Terra Nova* in heavy pack ice. (Herbert Ponting)

After studying all the evidence from previous voyages, they had entered the pack on the meridian of 178° W, believing they would have a good passage through, but now they were being rewarded by encountering the worst conditions imaginable, which meant more discomfort for the animals and for everybody. Scott bemoaned the state of the pack-ice before them and wrote:

> To understand the difficulty of the position you must appreciate what the pack is and how little is known of its movements. The pack in this part of the world consists (1) of the ice which has formed over the sea on the fringe of the Antarctic Continent during the last winter (2) of very heavy old ice floes which have broken out of bays and inlets during the previous summer but have not had time to get north before the winter sets in (3) of comparatively heavy ice formed over the Ross Sea early in the last winter and (4) of comparatively thin ice which has formed over parts of the Ross Sea in middle or toward the end of the last winter.

Undoubtedly throughout the winter all ice-sheets move and twist, tear apart and press up into ridges and thousands of bergs charge through these sheets, raising hummocks and lines of pressure and mixing things up; then of course where such rents are made in the winter, the sea freezes again forming a newer and thinner sheet.

With the coming of summer the northern edge of the sheet decays and the heavy ocean swell penetrates it, gradually breaking it into smaller and smaller fragments. Then the whole body moves to the north and the swell of the Ross Sea attacks the southern edge of the pack.

This makes it clear why at the northern and southern limits the pieces or ice-floes are comparatively small, whilst in the middle the floes may be two or three miles across; and why the pack may and does consist of various natures of ice-floes in extraordinary confusion. Further it will be understood why the belt grows narrower and the floes thinner and smaller as the summer advances ... We know that where thick pack may be found early in January, open water and a clear sea may be found in February and broadly that the later the date the easier the chance of getting through.

A ship going through the pack must either break through the floes, push them aside, or go round them, observing that she cannot push floes which are more than 200 or 300 yards across.

Whether or ship can get through or not depends on the thickness and nature of the ice, the size of the floes and the closeness with which they are packed together, as well as on her own power.

The situation of the main bodies of pack and the closeness with which the floes are packed depend almost entirely on the prevailing winds. One cannot tell what winds have prevailed before one's arrival; therefore one cannot know much about the situation or density. Within limits the density is changing from day to day and even from hour to hour; such changes depend on the wind, but it may not necessarily be a local wind, so that at times they seem almost mysterious. One sees the floes pressing closely against one another at a given time and an hour or two afterwards a gap of a foot or more may be seen between each. When the floes are pressed together it is difficult and sometimes impossible to force a way through, but when there is release of pressure the sum of many little gaps allows one to take a zig zag path.

They had entered the pack on 9 December, finally clearing it to gain the Ross Sea on 30 December, in Lat. 71°. Scott recorded that they had entered the pack with 342 tons of coal and left it with 281 tons, therefore expending

AB Mortimer McCarthy at the wheel. (Herbert Ponting)

61 tons, an average of 6 miles to the ton. Calculated in a straight line, they had covered 370 miles, an average of 18 miles a day. He had logged that they were steaming for nine out of twenty-one days with two long stops of five and four days when the boiler fires were drawn. (In a contrasting situation, it is worth mentioning here that when the author went south in January 2000, it took less than forty-eight hours to pass through the pack ice with the assistance of radar in a modern icebreaker of 12,500 tons and 25,000hp, at speeds varying between 9 and 12 knots.)

Christmas Day 1910 was celebrated with the ship trapped in the pack ice. The mess was gaily decorated and there was lusty singing of hymns at Service with a full attendance, all looking forward to a merry Christmas dinner that evening. One event celebrated at Christmas was the production of a family by Crean's rabbit:

She gave birth to 17, it is said, and Crean has given away 22! Scott wondered what would become of the parent or family ... at present they are warm and snug enough, tucked away in the fodder under the forecastle.

In 1841, Captain James Clark Ross RN, with two ships, HMS *Erebus* and HMS *Terror*, had found his way through the pack ice and into the Ross Sea which now bears his name. It was a remarkable feat, for neither ship had engine power, only sails. No doubt the course he had taken had been studied by Scott, but varying conditions at different times obviously dictate the outcome. Scott, clearly frustrated at their progress through the pack ice, wrote, 'The ponies are feeling the motion as we pitch in a short sharp sea, its damnable for them and disgusting for us'. In recognition of the ship, he recorded:

> The ship behaved splendidly no other ship not even the 'Discovery' would have come through so well ... As a result I have grown strangely attached to the 'Terra Nova,' as she bumped the floes with mighty shocks, crushing and grinding a way through some, twisting and turning to avoid others, she seemed like a living thing fighting a great fight. If only she had more economical engines she would be suitable in all respects.

Scott noted the antics of the Adélie penguins*. Now that the ship was in open water, more penguins were diving around the ship for food:

> The Adelie penguin on land or ice is almost wholly ludicrous. Whether sleeping, quarreling or playing, whether curious frightened or angry, its interest is continuously humorous, but the Adelie penguin in the water is another thing; as it darts to and fro a fathom or two below the surface as it leaps porpoise-like into the air or swims skimmingly over the rippling surface of a pool, it excites nothing but admiration. Its speed probably appears greater than it is, but the ability to twist and turn and the general control of movement is both beautiful and wonderful.

In the Ross Sea, they made their way through open water and with a few days' run they would be at Cape Crozier. It was now New Year's Day, 1911. They were under full sail and making 6 knots. Land had been sighted first on New Year's Eve and high mountains were visible to the westward, part of the Admiralty Range.

---

\* Named by the French explorer, Jules-Sébastien-César Dumont d'Urville, after his wife.

Lieutenant Evans in the crow's nest. (Herbert Ponting)

They were nearing Cape Crozier, where it was intended to set up base. This would give them a position for a shorter journey to the pole. Every few miles, Rennick took soundings, the depth varying from 357 fathoms to 180 fathoms as they closed on Cape Crozier. Oates took five of the ponies up on to the deck to clear out the stables which were between 2 and 3ft deep with manure.

Cape Crozier was reached on 3 January 1911, and the Great Ice Barrier appeared on the southern horizon. They steamed up close to the Barrier and then westward until reaching a little bay, where a boat was lowered to make a landing to choose a suitable base. Scott, Wilson and Lieutenant Evans and others went close inshore in the whaler but because of the swell they could not land. The decision was taken that no landing was possible and they returned to *Terra Nova*. The shore was surveyed from the ship, but ideas of a landing were abandoned, so course was set for McMurdo Sound.

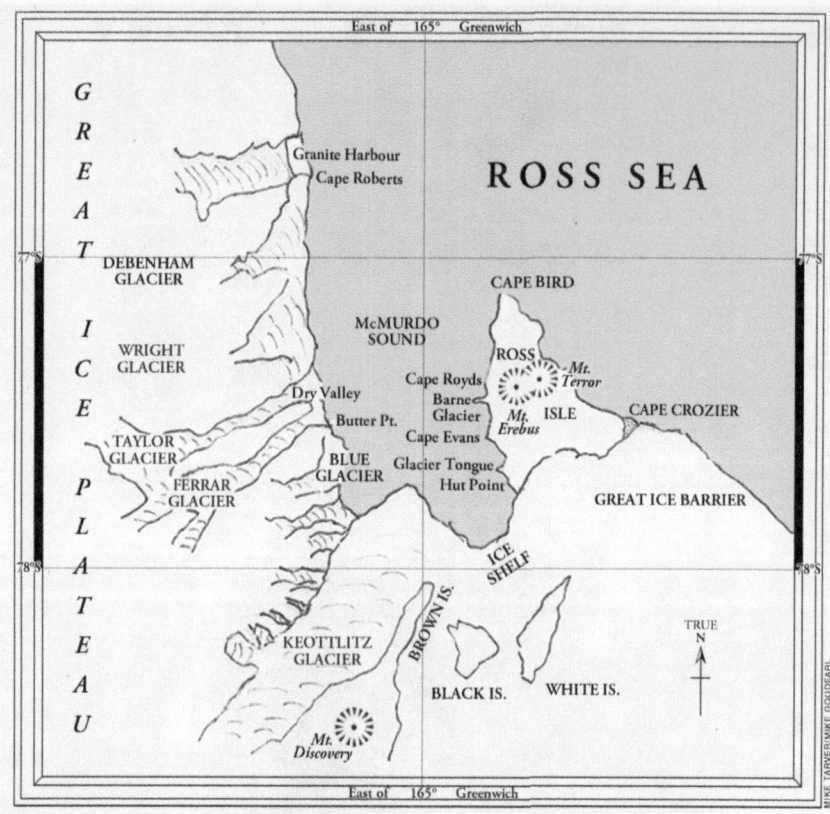

McMurdo Sound, Ross Sea. (Chart compiled by author)

## Arrival at McMurdo Sound and a Change of Command

From the crow's nest Lieutenant Evans worked the ship around Cape Bird into McMurdo Sound, with leadsmen taking soundings all the way. As they passed Cape Royds, they were able to spot Shackleton's hut through binoculars.

Scott, Lieutenant Evans and Wilson climbed the rigging and it was decided to make a landing on the beach at a cape, which they had previously called the Skuary and was at a position 15 miles north of Hut Point, the base of the 1901–04 *Discovery* expedition.

An ice-anchor was laid out and the three walked a mile to the cape, which all agreed would be ideal for a base. Scott renamed the Skuary as Cape Evans in honour of his second in command. The ship was unable to go further inshore because of the ice, making the distance between the ship and the site for the hut at Cape Evans about 2 miles.

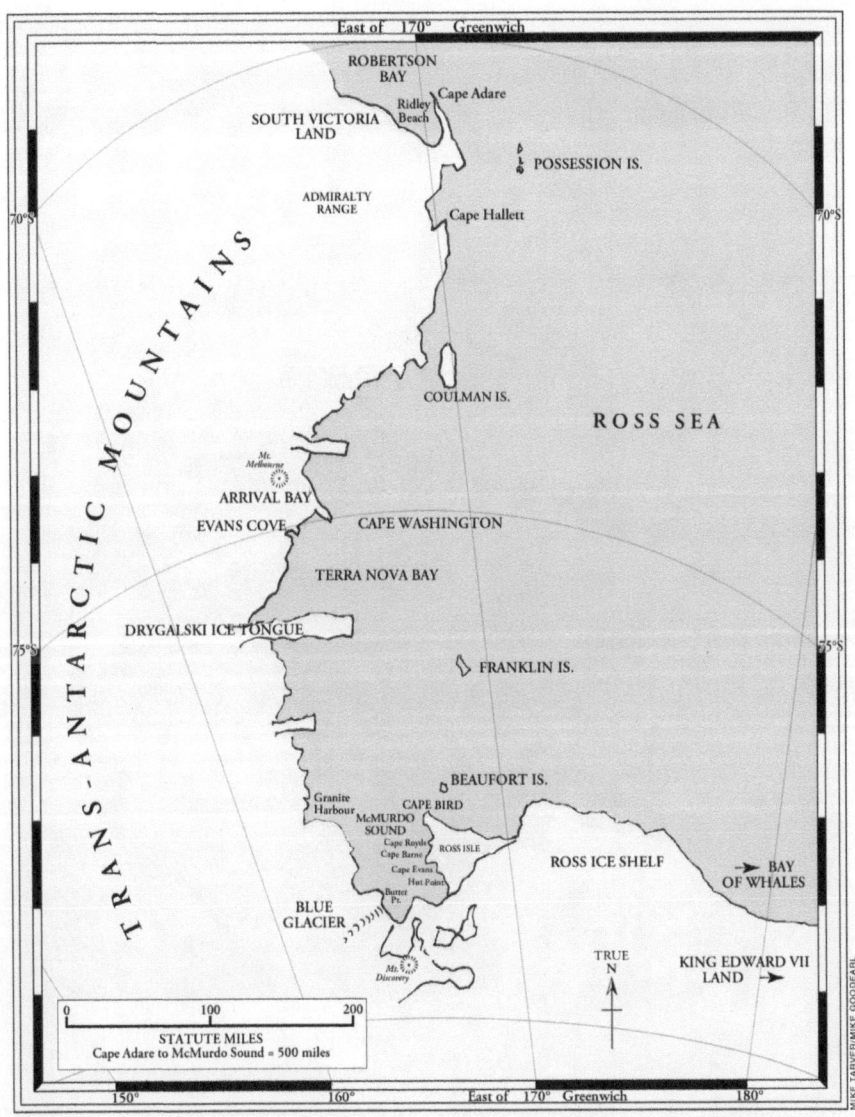

Ross Sea region of Antarctica, Cape Adare to McMurdo Sound. (Chart compiled by author)

Three parties were organised to unload and transport the equipment. Lieutenant Evans was to be in the shore party and command of *Terra Nova* was handed over to Lieutenant Pennell.

# 7

# First Role Complete

### Expedition Base Established, Scientific Parties Deployed

Work began immediately on preparing a foundation base for the hut, and tents were erected as temporary accommodation. Rennick came across the ice with the pianola all in pieces on sledges and set about putting it together. He tuned it and then played 'Home Sweet Home', to the delight of everyone.

All seventeen ponies were unloaded and once on the floe they kicked and rolled with joy. They were led to the beach and tied up where they could not eat the shingle sand. The dog teams were running light loads between ship and shore, but were constantly being interfered with by penguins, as each creature tried to identify the other with constant baiting, the dogs howling and straining at their leashes while the penguins squawked defiance and looked indifferently at the odd newcomers.

Two of the motor sledges were unloaded from the ship. The newer of the two safely landed was put to use and frequently towed loads of 2,500lb over the ice at 6 miles an hour. The older one hauled a ton and managed six double trips a day. Day, the motor operator, who with Priestley had been with Shackleton on the 1907–09 expedition, was not to blame for the motor sledges proving disappointing on the final polar journey.

They had landed in a period of fine weather. The warm glow of the sun with the keen invigorating cold of the air formed a combination of inexpressible health-giving satisfaction. 'The gold light on this wonderful scene of mountain and ice satisfied every claim of scenic magnificence', wrote

Artist's view of McMurdo Sound, Ross Island and surrounding area. (From *The Great White South* by Herbert Ponting)

Cherry-Garrard, 'No words of mine can convey the impressiveness of the wonderful panorama displayed to our eyes.'

On Sunday, 8 January, the third motor, one of two new ones, fell through the ice while being unloaded. Thomas Williamson, attempting to hold it with a rope, was pulled down through the ice but he was safely recovered. It was decided to move the ship along the ice edge to a firmer position.

They began to live in the hut on 18 January. It was warm inside and with the gramophone going everybody was happy. Scott was delighted with the hut, but he thought the word 'hut' was most inappropriate for such a palatial and comfortable dwelling place, in every respect the finest ever to be erected in the polar regions. It measured 50ft long by 25ft wide and 9ft to the eaves. 'What shall we call it?' he asked himself. He afterwards referred to it as 'our house'.

He recorded his appreciation of the men in all departments. He complimented and praised them in every way as work progressed, not only in the setting up of the base, but in the preparing for the depot-laying journey shortly to begin. The two scientific teams to east and west were assembling their equipment and the ship's crew were busily preparing for their departure to New Zealand.

On 17 January, the ice was breaking up between the ship and the shore and *Terra Nova* was obliged to get up steam. It all depended on wind direction.

With a southerly, she would bind to the ice with her anchors, but a change to a northerly wind would put her on a lee shore where the soundings run down to 3 fathoms at Cape Evans.

On 21 January, early in the morning, the ship was in difficulties. The ice was breaking up with a northerly swell and the northerly wind was increasing, putting the ship towards the lee shore. Pennell had steam up and the men were seeking to re-secure the ice anchors and get the ship head to wind and enable her to get away from the shore.

Later that day, when conditions allowed, *Terra Nova* returned, and just as she was turning to come alongside fast ice she struck a rock with only 12ft of water over it. This pinnacle lay within 20ft of a sounding of 11 fathoms. Pennell immediately sounded all round, shifted several tons of weight aft and had the engines going full speed astern with the crew running from side to side to roll the ship. Just as a shore party in a whaler came out to help, the ship moved astern and successfully floated off. Everybody cheered and the relief was enormous. She then anchored further off near Glacier Tongue. (When *Terra Nova* was dry docked on her return to New Zealand it was found that a furrow several yards long had been gouged into her bottom, but she had not sprung a leak.)

These were more than anxious moments, for there was no relief ship arranged. Had *Terra Nova* foundered, sixty-five men would have been marooned for an unspecified time. It did not bear thinking about. Lieutenant Evans and Wilson went aboard to console Pennell and brought him back to Cape Evans for tea. The event was described to Scott, who was sympathetic to Pennell and related his own story of the near miss they had experienced when departing from Hut Point in 1904 in the *Discovery*. In his journal, Scott refers to the incident and Pennell's visit, which he seems to have taken calmly. He recorded, 'From Pennell down, there is not a man who has not done his duty nobly during the past weeks. Pennell has been over to tell me about it tonight; I think I like him more every day.'

On 26 January, letters for home were put aboard and *Terra Nova* prepared to depart to deploy the scientific parties before returning to New Zealand. Lieutenant Victor Campbell and his eastern scientific party, consisting of Raymond Priestley, Murray Levick, George Abbott, Harry Dickason and Frank Browning, were ready to go aboard to be taken to King Edward VII Land, as were the members of the first western geological party, consisting of Griffith Taylor, Charles Wright, Frank Debenham, and PO Edgar Evans, who were to be landed across McMurdo Sound at Butter Point. Ponting took photographs, then Scott went to the ship with a dog team and bid them farewell.

The Hut at Cape Evans with Cape Barne in the distance, British Antarctic Expedition 1910–13. (Courtesy Hon. Edward Broke Evans)

Work was beginning to prepare for the southern journey. The first task for the shore party at Cape Evans was to lay depots south from Hut Point over the Great Ice Barrier as far as Lat. 80° S. This work would have to be completed within weeks and before the coming of the colder weather and the onset of their first Antarctic winter. The attempt to reach the South Pole would be made toward the end of the year after the sun returned.

As the depot-laying journey was beginning, *Terra Nova* departed Cape Evans on 27 January 1911, with the two scientific parties aboard to be dropped off at their designated points. The western party were, as planned, put off at Granite Harbour and then passage was made north to disembark the eastern party on the way to New Zealand – but it didn't quite happen like that.

Scott's written instructions to Campbell already took note of the fact that there might be a difficulty in finding a suitable landing place and conditions in which to operate to discover the nature and extent of King Edward VII Land. On the journey south, attempts had already been made from *Terra Nova* to find a suitable landing area, but this was not successful, and the outcome of this eastern party attempt was speculative. It was arranged that should they land, the ship would stand by for three days while they established themselves. Should they be unable to land at King Edward VII Land, it was provided in the instructions that they could land in the Robertson Bay area at Cape Adare and explore westward of Cape North.

## A Surprise in the Bay of Whales

By 31 January 1911, as *Terra Nova* passed along the coast beyond Cape Crozier, they entered a number of bays and on reaching the Bay of Whales late on 3 February they came across a ship fast against the sea ice. It was immediately recognised by Campbell as the Norwegian ship *Fram*.

Campbell takes up the story:

> Standing in, we made fast a little way ahead of her and hoisted our colours, she answering with a Norwegian ensign. There was no doubt it was Captain Amundsen. Pennell and I immediately went on board and saw Lieutenant Neilsen, who was in command. He told us that Amundsen was up at the camp about 3 miles in over the sea ice, but would be back about 9 o'clock and accordingly soon after nine I returned on board and saw Amundsen, who told me his plans. He had been there since January 4th after a good passage, having been held up only 4 days in the pack. He had intended wintering at Balloon Bight, but on finding that it had gone, had fixed on the Bay of Whales as the best place.
>
> He asked me to come up and see his camp, so Pennell and I went up and found he had erected his hut on the Barrier about 3 miles from the coast. The camp presented a workman like appearance, with a good sized hut, it contained a kitchen and living room with a double tier of bunks round the walls, while outside several tents were up and 116 fine Greenland dogs picketed round. His party, besides himself, consisted of Johansen, who was with Dr. Nansen in his famous sledge journey of '97 and seven others. After coffee and a walk round the camp, Amundsen and two others returned with us and had lunch in the 'Terra Nova.' We left early in the afternoon and after dredging and sounding in the bay, proceeded west along the Barrier of which there still remained nearly 100 miles we had not seen.

The intention of setting up a base for the party in King Edward VII Land was abandoned and the secondary option, that of a landing at Cape Adare was decided upon. However, it was important that the news of Amundsen's pres-

Roald Amundsen's ship *Fram* in the Bay of Whales with *Terra Nova* standing off. (*The Amundsen Photographs*, Hodder & Stoughton, 1987)

ence should be communicated to the main party and so *Terra Nova* returned at once to McMurdo Sound, arriving on 8 February.

Two ponies not now required for the party's activities were landed at Cape Evans. The ship then anchored off Glacier Tongue from where Campbell, Priestley and Abbott made their way to Hut Point where the depot-laying party would call on the return from the Barrier. Here, a letter was left for Scott with the news of Amundsen's presence in the Bay of Whales.

On returning to Glacier Tongue, an accident happened when they were moored alongside watering the ship. While Abbott was stepping ashore, a large piece of ice broke away with him and he fell between the ship and the ice edge. Abbott was not hurt and was quickly pulled back aboard, but he did get a ducking.

On the evening of 9 February 1911, *Terra Nova* made her way north with a south-easterly wind and in thick snow. By 12 February, they were off Cape Adare, but weather conditions caused them to stand off in a gale and thick snow, Cape Adare bearing north-west at a distance of 20 miles.

The problem with these unscheduled movements by *Terra Nova* was the extra use of coal, which could hardly be spared. Weather conditions were preventing them from approaching their objective and by 15 February they were standing many miles off Cape Adare, which now lay to the south-west. The coal question was becoming serious, for if they were to be delayed much longer, they would not be able to land Campbell's party, as Pennell had to keep enough coal to get back to New Zealand.

On 16 February, they were able to set lower topsails to a south-easterly wind and by the early hours of 17th they were within 2 miles of the coast. They entered Robertson Bay and made for Ridley Beach on the west side of Cape Adare, where Borchgrevink's party with the *Southern Cross* had wintered in 1900. For Campbell, it was not the ideal choice to set up a base because deeply sloping cliffs came right down to the sea and made access inland difficult from the bay.

The main consideration remained the amount of coal in the ship. Pennell could not be delayed further. Stores and equipment were landed, and the base hut was built by Francis Davies, the ship's carpenter in forty-eight hours non-stop, after which *Terra Nova* hurriedly departed for New Zealand on 20 February.

In the rush, only two hammers were landed, leaving the party now known as the northern party, with two of Priestley's geological hammers, described by Campbell as heavy, square-headed implements designed to chip, resulting in their getting mangled fingers. There was more bad luck when they examined fifteen carcases of frozen mutton left by the ship only to find them covered in green mould. These were buried in the ice, much to the delight of the skuas. Penguins and seals however, were plentiful in the area and were a good source of fresh meat.

Francis Davies, Leading Shipwright RN, made a considerable contribution to the expedition. On board *Terra Nova*, he was very much involved in emergency work in the engine room during the storm in the Southern Ocean. Prior to departure from Britain, his skills as a shipwright came to the fore in preparing the ship for the expedition and afterwards in building the huts at Cape Evans and at Ridley Beach, Cape Adare for the northern party. He also made the cross erected at Observation Hill in memory of the five expedition members who died on the return journey from the South Pole.

The Terra Nova Agreement and Account of Crew shows that Davies was born at Plymouth and was 25 years of age. He joined the British Antarctic Expedition from HMS *Vanguard* on 30 May 1910 and signed on at Poplar, London. His home address is shown as Nicholls Farm, Plympton, Devon.

## The SS Terra Nova

*Terra Nova* in the Ross Sea. (Herbert Ponting)

In *The Worst Journey in the World*, Apsley Cherry-Garrard wrote of him:

> Davies the carpenter deserves much credit. He was leading shipwright in the navy, always willing and bright and with a very thorough knowledge of his job. I have seen him called up hour after hour, day and night on the ship when the pumps were choked by the coal balls which formed in the bilges and he always arrived with a smile on his face. Altogether he was one of our most useful men.

The following extract is taken from the biography of Davies in *British Polar Exploration and Research 1818–1999* by Neville Poulson and John Myres:

> After the Antarctic, Davies returned to service in the Royal Navy and during the First World War he served aboard HMS *Blanche* with the Grand Fleet and on mine laying. In 1918 he served in HMS *Exmouth* and in 1919 he went to Archangel in Russia where he was placed in charge of docking and repairs to all shipping operating in the White Sea and Severnaya Dvina River. In 1919–20 he served in HMS *Sandhurst* during operations in the Baltic after which he retired voluntarily. In August, 1932, he passed the Board of Trade examination for Certificate of Competency as a Master (Steam Ship-Foreign Going).

During the period 1927–34 he served in the Royal Research Ships *Discovery II* and *William Scorsby* which were engaged on scientific work in the Southern Ocean regions. He volunteered for service in the Second World War and was granted a Temporary Commission as a Lieutenant, RNVR. In 1940 he was appointed to HMS *Victory III* for special service, and at Harstad, Norway he served with the naval detachment until evacuation. He was commended for his services in Norway. Later in 1940 he was appointed to Boom defences at the Wash and then as Temporary Lieutenant Commander, RNVR, Boom Defence Officer, Grimsby.

Sometime after the Second World War, Davies typed a 100-page account of his experiences with the British Antarctic Expedition, 1910–13 entitled 'With Scott before the Mast' by Rudolf. The original work is in the Marlborough Museum, Christchurch, New Zealand. There is a copy on microfilm at the Scott Polar Research Institute, University of Cambridge, but due to technical difficulties it cannot be transcribed. I would have liked to have included extracts from Davies's recollections with this story, but his narrative must now be left for others to retell on his behalf.

## Scientific Parties Landed and the Return to New Zealand

*Terra Nova* was now on passage to New Zealand. Base Camp for the shore party was established at Cape Evans and a start had been made to lay depots south in preparation for the main southern polar journey to begin in October or November 1911. Two scientific parties had been landed, one for exploration of a region on the other side of McMurdo Sound and the other, the northern party at Cape Adare, Victoria Land, to explore that region. Added to this, Scott had been warned that Amundsen was in the Bay of Whales.

*Terra Nova* would next return to McMurdo Sound in January 1912. On her return, it was planned she would take the northern party aboard at Cape Adare, with instructions to land them further south at Evans Cove for a four-week expedition in that region. Also, the ship was to take aboard the second western geological party in the region of Granite Harbour, return them to Cape Evans and relieve the main expedition. This voyage would include delivering mules and fresh dogs brought from New Zealand, putting ashore staff as changes required and taking off those members of the expedition who were to return as arranged after serving for one winter.

Base camp established, *Terra Nova* departs for New Zealand on 27 January 1911. (Herbert Ponting)

The shore party of the British Antarctic Expedition, 1910 had now been established at their base. The expedition ship had successfully carried out her first tasks. A busy programme for the ship still lay ahead, for now there were hydrographic surveys to be carried out in New Zealand waters by arrangement with that government.

The programme of relief duties on the ship's later return to the Antarctic was to turn out to be difficult and complicated. Sea ice conditions and the usual unpredictable and vicious weather of the Antarctic continent would always present a challenge. It is not intended to include a full account of the British Antarctic Expedition, 1910–13 in the story of the *Terra Nova*, but a summary of principal events which occurred during that expedition is set out later in this narrative.

# 8

# NEW ZEALAND REFIT AND HYDROGRAPHIC SURVEYS

## Summary of Expedition Relief Voyages

Terra Nova had left Cardiff on 15 June 1910 and via Madeira, South Trinidad Island, South Africa, Australia and New Zealand, she arrived in Antarctica. The expedition reached Cape Evans on 4 January 1911, where Base Camp was established. *Terra Nova* had brought the expedition to McMurdo Sound and her first task was complete.

The captain of *Terra Nova* and deputy leader of the expedition, Lieutenant E.R.G.R. Evans, was in the shore party, thus command of the ship had now passed to Lieutenant Harry L.L. Pennell for the hydrographic survey work around New Zealand and to bring the ship back to relieve the expedition after its first Antarctic winter.

It will be remembered that on Scott's first expedition, *Discovery* had remained in McMurdo Sound during the two Antarctic winters and the decision to do so or not had been left to the expedition leader, as captain of his ship. After the *Discovery* experience, it had been decided that *Terra Nova* would not winter in the Antarctic but would return to Lyttelton in New Zealand. Lessons had been learned by experience; a luxury not previously enjoyed in 1901–04 on the then unknown continent.

(An incident subsequent to this story occurred in May 1915, when *Aurora*, expedition ship to the Ross Sea party of Shackleton's Trans-Antarctic Expedition of 1914, was blown north from her anchorage in McMurdo Sound. She was unable to return because of bad weather and ice damage to the ship, which left ten men abandoned on the continent for two winters. *Aurora* drifted through the pack ice for ten months with eighteen men aboard before getting back to New Zealand. Vicious winds and conditions promised little but trouble for expedition ships in exposed anchorages. In this case, it left explorers inadequately equipped for overwintering on the continent.)

After disembarking the first scientific party, discovering the *Fram* in the Bay of Whales and returning to inform Scott, *Terra Nova*, under the command of Lieutenant H.L.L. Pennell eventually left McMurdo Sound on 9 February 1911. Finally, she deployed Campbell's party at Cape Adare and by 20 February she was bound for New Zealand.

At Lyttelton she was dry docked and refitted, then carried out a programme of hydrographic survey work for the New Zealand Government.

Departing Lyttelton on 15 December 1911, *Terra Nova* returned to the Antarctic to relieve the expedition and arrived at McMurdo Sound on 4 February 1912. She departed again for New Zealand on 3 March, where she arrived on 1 April, 1912. Further hydrographic survey work was conducted.

On the ship's return to New Zealand on 3 March 1912, *Terra Nova* had aboard Lieutenant Evans, who had not long returned from the southern polar journey and was suffering from scurvy, his life having been saved by William Lashly and Tom Crean, who effected his return to base. He returned on *Terra Nova*, in his own words, 'a physical wreck'. He went back to Britain, where he recuperated, before returning to New Zealand later that year to rejoin the ship and resume command as captain for the last voyage to relieve the expedition in 1913.)

On 14 December 1912, *Terra Nova* once again left Lyttelton to relieve the expedition for the last time, on this occasion under the command of Commander Evans (in the meantime promoted from lieutenant). The ship reached Cape Evans on 18 January 1913.

The expedition eventually departed from Antarctica on 26 January 1913 and returned to Lyttelton on 12 February, where Commander Evans handed over command to Lieutenant Pennell. *Terra Nova* departed Lyttelton on 13 March under the command of Lieutenant Pennell for the final voyage to Cardiff where she arrived on 14 June 1913.

An account of these intervening voyages by the ship and the return to Cardiff was jointly prepared by Commander Evans and Commander Pennell, under the title of *Voyages of the Terra Nova* and appears in Volume II of *Scott's Last Expedition* by Leonard Huxley, published in 1913. The following chapters describe the many events that took place during the voyages between Antarctica and New Zealand and link up with the main story.

## First Return Voyage to New Zealand – 'The Pumps Again!'

After landing the northern party and equipment at Cape Adare, *Terra Nova* set course for New Zealand on 20 February 1911. Whales kept close to the ship till noon on 4 March when steam was raised, and good progress was made. They passed 10 miles west of Young Island of the Balleny Group, but the islands were covered in clouds and no useful bearings could be taken. By 8 March, they cleared the last of the pack and were pleased to be joined by the sooty albatross and many other deep-sea birds.

For the next two weeks, it was a struggle to keep their course as the wind held to the north, which made for a stiff sail. Pennell recorded:

> To the seaman of the present day used to iron ships, it is a never-failing source of surprise and delight to see a wooden ship in a heavy sea. How nicely she rides the waves, like a living being, instead of behaving like a half-submerged rock.

The albatross and other deep-sea birds gave them much pleasure and south of Lat. 60°, the pretty hourglass dolphin (first noticed by Dr Wilson in the *Discovery*) was often around the ship.

On 22 March, 90 miles south of Macquarie Island, the long-hoped for fair wind came at last and held until they were off Stewart Island, New Zealand.

The pumps had been a nuisance throughout and during a gale on 24 February, the trouble came to a head with the ship heeling at between 40 to 45 degrees and jumping about considerably. During such a heel, coal and ashes tipped over and blocked the pump-well in the engine room. The chief engineer, William Williams, with great difficulty keeping a perforated enamel jug on the end of the suction, stopped the pump every two or three minutes as the suction choked, then removed and cleared the jug, replaced it and then restarted the pump. This process had to be kept up the whole time the hand pumps were being seen to. To achieve this, Williams had to lie flat on the boiler-room plates and as the ship listed to starboard, stretch right down with his head below the plates and clear as much coal away from the suction as possible. He had to get away before the water surged back and he wasn't always in

Steaming through the pack ice. (Courtesy Hon. Edward Broke Evans)

time! Meanwhile, others were working on modifications to the pumps. The whole job took eight hours to complete.

They made Lyttelton on 1 April 1911. There, refitting the pumps and everything connected with them was carried out. They gave no further trouble.

Modifications to improve *Terra Nova*'s power and performance in ice were made to the ship's engine by Mr Dickson of Lyttelton, at the suggestion of William Williams, the chief engineer. An increase in forward revolutions from 60rpm to 89rpm was obtained, retaining 60rpm in reverse. The ship lay at Lyttelton for three months and underwent a thorough refit. During this time, tasks were allocated for the winter cruise to come, while Mr Joseph Kinsey helped the expedition in every way he could.

## Winter Cruise and Hydrographic Survey Work (10 July–10 October 1911)

The scientific side of the ship's work was undertaken as follows:

Biological research – Dennis G. Lillie
Zoological log – Lieutenant Wilfred Bruce

Operating the Lucas deep-sea sounding machine. (Photo by AB James Skelton)

Magnetic & current log – Lieutenant Harry Pennell
Soundings & chart preparation – Lieutenant Henry E. de P. Rennick
Refit – Lieutenant Wilfred Bruce

The ship left Lyttelton for the winter cruise on 10 July 1911 and returned on 10 October 1911. The main objective was to carry out a survey round Three Kings' Islands at the extreme north of New Zealand. Here, the swell from the Tasman Sea to the west met that from the Pacific to the east, causing a confused sea at all times. The routine was to make soundings at day and put the plankton nets out at night. Rennick and Williams ingeniously adapted a motorboat engine lent by Mr Kinsey to work the Lucas sounding machine, improving its working efficiency.

*Terra Nova* was fitted with a Lucas deep-sea sounding machine, and some explanation of its origins and functions is needed. Interest in scientific exploration of the depths did not come until early in the nineteenth century. Before this, Britain's interest at sea had been only in warfare, protecting her growing Empire and expanding trade routes.

It could be said that scientific interest really began with Charles Darwin and the voyage of *Beagle* (1831–36), commanded by Robert Fitzroy, and later by the round-the-world voyage of HMS *Challenger* (1872–76). On 16 February 1874, HMS *Challenger* was the first steam ship to go south and cross the Antarctic Circle at Lat. 66° 33' S, standing on for 10 miles.

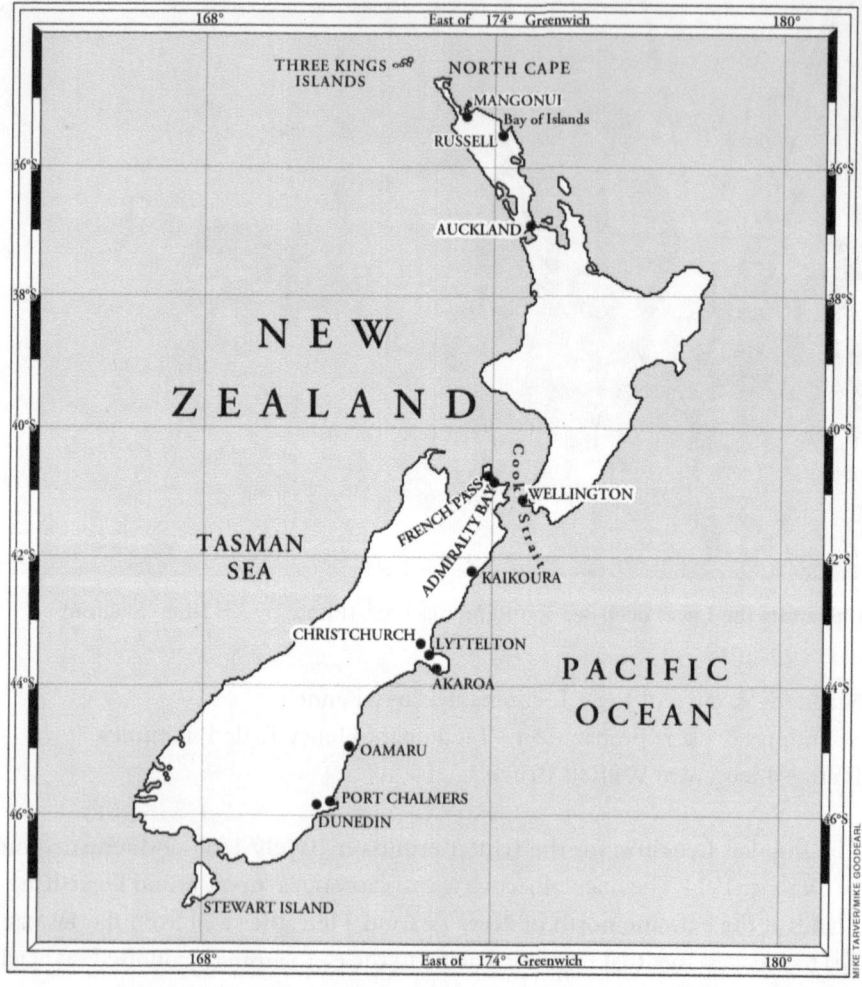

At Edinburgh University, Professor Edward Forbes (1815–54) had propounded the theory that life at sea did not exist below the depth of 300 fathoms. He gave the name to these depths as the Azoic Zone, i.e. a zone where life could not exist. His theory was soon to be challenged and proved wrong.

Captain George Strong Nares RN had initially commanded HMS *Challenger* on what was really the first scientific round-the-world voyage. At Hong Kong, he handed over command of HMS *Challenger* to Captain Frank Turle Thomson. Nares was recalled to command the British Arctic Expedition, 1875–76 in HMS *Alert* accompanied by HMS *Discovery* (formerly the Dundee whaler, *Bloodhound*, built by Alexander Stephen & Sons in 1872).

This Arctic expedition returned prematurely because of a serious outbreak of scurvy, for which Nares was blamed. After a committee of enquiry which was inconclusive, Nares was again given command of HMS *Alert*, this time for surveying off South America and in the Pacific. It was on this project that his ship was provided with a Lucas deep-sea sounding machine for its first naval trial. The machine was developed by Francis Lucas, chief engineer of the Telegraph Construction & Maintenance Company and by Sir George H. Richards, formerly Hydrographer of the Navy, now the company's managing director.

A photograph shows the Lucas machine, fitted starboard side, forward of *Terra Nova*. For a more detailed description of its functions, I refer to *The Home of the Blizzard* by Sir Douglas Mawson, the Australian Antarctic explorer, who describes the machine and its functions when fitted to his exploration ship, *Aurora*:

> The Lucas deep-sea sounding machine, designed for ascertaining depths up to 6,000 fathoms, was mounted on the port side [starboard side on the *Terra Nova*] of the fo'c'sle head. In this machine was mounted a fine strand of steel wire. As it is paid out, weighted heavily at the end, it is made to pass over a wheel which measures and records the number of fathoms. The action of the sinker striking the bottom causes a spring brake to stop the reel or wire from further unwinding; at the same time the sinker automatically disengages, for it is more economical to use a fresh piece of cast-iron each time than laboriously to haul the weight from the bottom of the sea. When in addition to ascertaining the depth; samples of the bottom are required, other devices, which automatically secure specimens of the mud or rock as they strike the bottom, are attached to the end of the wire.
>
> There was also a larger mechanism operating with five-ply wire designed for obtaining serial temperatures and water samples. So there were, altogether, three sounding machines fitted on the Aurora besides the small Kelvin instrument astern which was used mainly for purposes of navigation. The arduous work of winding in the wire after each sounding was eliminated by the application of a small steam-driven engine, which had been specially designed for the Scotia and was kindly lent to us by Dr W.S. Bruce.

On the *Terra Nova* voyage, sounding could not be performed if the wind was easterly. They could only heave to and drift. If westerly, they found a good anchorage inside North Cape (the extreme north-east point of New Zealand), where they were grateful for quiet days aboard and where engineering work could be carried out. Occasional visits were paid to Mangonui, on the east coast to take on provisions.

On 22 September 1911, they sailed south again. On the way they called at Russell, Bay of Islands, for provisions and to collect mail. Pennell describes the area as 'very pretty, historical and interesting'. Lillie and another walked over the peninsula to the tiny Bay of Wangamumu where a small whaling station operated.

The ship arrived off Kaikoura on 8 October and they spent two days sounding before putting into Lyttelton. They remained there for the next two months, Rennick putting finishing touches to his chart of the Three Kings Islands.

A summary of the biological work conducted aboard ship between 1910 and 1913 is the subject of a separate report by Dennis G. Lillie in *Scott's Last Expedition* by Leonard Huxley and appears in a later chapter.

Waiting to be brought aboard ship at Lyttelton were the seven mules given by the Indian Government and the fourteen Siberian dogs from Vladivostok. In India, the mules had been exercised in rocking boxes to help develop their muscles for travel in the ship. Stables were built on the foremost side of the ice-house so that they were in the open air. James Dennistoun now joined the expedition to take care of the mules.

The programme for the return voyage south, according to the outline given by Captain Scott in his sailing orders, was to be as follows:

1 Pick up Campbell and the northern party about January 1st at Cape Adare.
2 Reland them in the vicinity of Evans Cove, Wood Bay.
3 Relieve the 2nd geological party (Griffith Taylor) about January 15th at Granite Harbour.
4 Land mules, dogs and stores at Cape Evans.
5 Lay out various depots according to the orders to be received at the Hut in readiness for the next season's work.
6 Consistently with carrying out the above, make biological collections, take soundings and carry out other scientific work to as large an extent as possible.

# 9

# RETURN TO ANTARCTICA – FIRST RELIEF VOYAGE

## McMurdo Sound Relief and Attempted Relief of Scientific Parties

Terra Nova departed Lyttelton on 15 December 1911 with all relief stores aboard, also the seven mules and the fourteen dogs. On the whole, it was a good smooth passage under steam to the pack ice and they were able to move the mules and clean out the stables.

On the Sunday before Christmas, just as they were going to lunch, N*****, the ship's cat, fell overboard. He had been baiting the dogs on the poop, had got too close to one, jumped to avoid the dog and went over the side. Fortunately it was a calm day, the sea boat was lowered and N*****, who swam pluckily, was picked up. The ship was on her course again twelve minutes after the incident. He was quite benumbed with the cold but was taken down to the engine room and well dried, given a little brandy to drink and by the evening was all right again.

Cherry-Garrard wrote of N*****:

This cat became a well known and much photographed member of the *Terra Nova*'s crew. He is said to have imitated the Romans of old, being a greedy beast, by having eaten as much seal blubber as he could hold, made himself sick and gone back and resumed his meal. He had most beautiful fur.

But N*****'s luck was to run out on the return voyage.

N*****, the ship's cat. (Herbert Ponting)

The first iceberg was passed on Christmas Day and the first of the pack ice on Boxing Day, but by the 28th they were held up in heavy pack ice. Taking advantage of the hold-up, they exercised the mules round the deck and ran the dogs on the ice. At times, Rennick took soundings and Lillie was able to lower plankton nets.

On 31 December, they resumed good progress with steam and reached Cape Adare on 3 January 1912. The following day, they were able to take Campbell and his party aboard with their collections and equipment. All were in good spirits, they had been at Cape Adare since 20 February 1911.

'The tidal streams are strong and the pack ice heavy at Robertson Bay,' recorded Pennell, and he was pleased to be away quickly. They made their way south to Arrival Bay, 6 miles north of Evans Cove. With gear and a month's depot, they put Campbell and his party ashore, as arranged, on 8 January 1912.

After negotiating the pack ice, the ship arrived off Beaufort Island on 12 January. There was heavy ice in the direction of Granite Harbour and across the entrance to McMurdo Sound. They used the enforced delay to swing the ship for deviation in magnetic readings, to collect plankton, and to record wildlife observations. Whenever there appeared to be a change in the ice conditions, they would try to steer the ship towards Granite Harbour or Cape Evans, but without avail.

On 4 February, they were able to secure alongside fast ice off Cape Barne. Atkinson, Meares and Simpson came out with dog teams from Cape Evans and discussed the ice conditions but left quickly, taking the mail with them.

The sea ice went out with a gale on 6 February and they were able to secure the ship to fast ice off Cape Evans, but still they were more than 2 miles out. The mules and the dogs were landed and stores were sledged to the base camp. Unloading took place steadily until 14 February, when a further gale sent out more sea ice, putting the ship closer to Cape Evans.

On 19 February, still with some stores aboard, they took the ship over to Butter Point to locate the western party and found a note from Griffith Taylor stating that they had gone south to Blue Glacier. There, *Terra Nova* boarded them. They had anticipated the difficulty of the ship entering Granite Harbour. All the party were well, but could not be landed at Cape Evans, so the ship went north again intending to pick up Campbell's party at Terra Nova Bay. Meanwhile, Mather, a skilled taxidermist, was kept busy skinning specimens of wildlife.

It took time to negotiate the pack ice around the area of the Drygalski Barrier. For two days, the ship was in a storm with little visibility. Nothing could be done but to remain under easy steam, avoid floes and look out for icebergs.

After two days of these conditions, Mount Melbourne appeared with great beauty, but by 21 February there was still no prospect of reaching Campbell's party. In these difficult conditions, *Terra Nova* regained open water and on 23 February, made another unsuccessful attempt to relieve the party. It took four hours to turn the ship and another twenty-six hours to get out of the pack ice. The ship then made for Cape Evans to land Griffith Taylor's party. They were going to have to devote some time endeavouring to relieve Campbell.

On 25 February, they were able to anchor close in at Cape Evans, all the fast ice having gone out since they were last there. It was then learned that Lieutenant Evans from the returned polar party was at Hut Point and seriously ill and should be taken aboard as soon as possible. A gale was blowing and for three days it was impossible to come in.

They were able to secure alongside fast ice just north of Hut Point on 28 February, and Atkinson and his party were able to bring Lieutenant Evans aboard. Atkinson remained on the ship to care for him, as the attention of a doctor was necessary for a few days. On return to Cape Evans, all the remainder of the stores, about 19 tons, was unloaded in calm and beautiful weather.

Because of his condition, Lieutenant Evans remained aboard to return to New Zealand. The ship immediately left for Terra Nova Bay but found that conditions had not improved, the ship running up and down and trying to get through a weak place in the ice. On one occasion she made it through, only to still be 35 miles off. Entries in the ship's log show in great detail all the attempts to relieve Campbell.

On 3 March, the ship again returned briefly to Cape Evans, to embark those returning home and to land Atkinson and Keohane. The ship then ran to Terra Nova Bay again, only to find the conditions worse than before.

## Second Return Voyage to New Zealand and More Survey Duties

Finally, on 7 March 1912, they reluctantly decided that the ship could not reach Campbell's party that season, so she was turned north.

(Nearly ninety years later, in January 2000, the author was landed by small craft at Terra Nova Bay with a party from the powerful Russian icebreaker, *Kapitan Khlebnikov*. We were ashore for less than two hours and in that time, weather conditions changed, ice was blown into the bay, and conditions prevented our being taken off by sea. We were fortunate in that the ship's helicopters were on hand to return us to the ship.)

The story of Campbell's party, Levick, Priestley, Dickason, Abbott and Browning, from their landing on 8 January 1912 and their survival in an ice cave until their eventual return to Cape Evans on 6 November 1912, must be one of the greatest of survival stories ever. Seeing the difficulties and unlikelihood of being relieved by ship, they constructed an ice shelter and endured a second winter, finally negotiating their own escape and return to Cape Evans.

They began their return to Cape Evans on 30 September 1912 by sledging round the coast, arriving at Base Camp on 6 November 1912. Not only had they spent nearly a year in a hut on a remote beach at Cape Adare, but this was then followed by nine months in a cave built by themselves at Evans Cove, after being landed there by *Terra Nova* originally only for four weeks. Compare this with the relative comfort of the base hut at Cape Evans and below decks of *Terra Nova*, both of which they must have yearned for. Their survival through two Antarctic winters says much for the competent leadership of Lieutenant Campbell and for the services-trained discipline of the men.

On 15 and 16 March 1912, the ship passed the north-east side of the Balleny Islands and when the fog cleared on the 16th, 'we caught a glimpse of Buckle Island, part of a snow capped mountain with the sun on it, a rarely beautiful sight, appearing to be quite detached from the earth itself'. On the 25th they suffered a storm, the most severe encountered during her whole commission, recorded Pennell:

> It is a wonderful sight to see a comparatively small ship in a storm, particularly at night; the marvellous way she rides over the waves that look as if they must break on board, together with the dense darkness in the heavy squalls, relieved only by the white crests of the waves as they break, it is a sight that makes up for a considerable amount of discomfort.

Ponting described the seas in *The Great White South*:

> The ship was now in ballast, and light, and it was magnificent to watch her fighting the mountainous seas which seemed, at times, as though inevitably they must engulf her. One minute she would be in an ocean valley, with waves ahead and astern higher than her main top, the next she would be on the summit of one of these watery peaks. The storm provided a thrilling subject for the last phase of our adventure that I recorded in moving-pictures.

It would seem that during this big gale N*****, the ship's cat, was lost one dark night. He often went aloft with the men of his own accord, 'This night he was seen on the main lower topsail yard, higher than which he never would go. He disappeared in a big squall, probably because the yard was covered with ice.'

The storm was followed by two days' calm, which allowed Ponting, who was returning home as his main work was complete, to kinematograph the birds close to the ship's stern. Off the coast of New Zealand, a school of sperm whales was seen to follow the ship for some time.

Ponting recorded:

> A day's steam from land, we sighted a school of Sperm whales lazily spouting in the sunlit waves. They excited immense interest aboard, for these now rarely seen leviathans of the ocean are quite different from any other whales. Their heads, which are enormous, are shaped like the bow of a battleship. They spout diagonally forward, not vertically as do other whales, and the

*Terra Nova* at Akaroa Harbour, 1 April 1912. It was on this date that the ship had returned from the Antarctic after her first relief voyage with members of the expedition aboard who were returning home having served their agreed tour of one winter. Also returning home was Lieutenant Teddy Evans, Scott's deputy leader, who was seriously ill with scurvy. (I am grateful to Mr Clive Goodenough for this photograph, formerly of Christchurch, now living in Melbourne. His maternal grandfather, Charles Johnston, was employed by Sir Joseph Kinsey in the capacity of marine superintendent for Kinsey's Shipping Company. Charles Johnston was responsible for provisioning the ship and sailed out to the Heads of the harbour on *Terra Nova* when she departed Lyttelton on 25 November 1910, returning by pilot boat. On the final return of the expedition to Lyttelton in February 1913, the ship's company presented Charles Johnston with a sledge, much used by his sons on the hills around Lyttelton.)

spiracle is in the front, instead of in the middle of the head. For some time the ship was manoeuvred in the hope that we night get near enough for me to secure some kinematograph records; but the great creatures were too wary, and kept a good quarter-of-a mile out of range.

At daybreak on 1 April 1912, the ship entered Akaroa Harbour to despatch telegrams and here they learned of Amundsen's success in attaining the South Pole. On 3 April, they berthed alongside at Lyttelton once again to receive New Zealand's hospitality. It had been a hard season for the ship and Pennell recorded his appreciation of the crew, in particular the engine room department.

## Another Refit in New Zealand and a Tragedy at Admiralty Bay

After an improvement in his health, Lieutenant Evans, together with Drake, went home to the UK and there were some crew changes. Laying up and refitting the ship was carried out and arrangements were made for the ship's party to survey Admiralty Bay in the north of South Island. Work was accomplished from motor launches, again fitted with Lucas sounding machines, while the crew were accommodated at Kairanga House, French Pass.

The work lasted from 10 June to 15 October 1912. Lieutenant Pennell's diary records that the surveying party numbered fourteen, i.e. three lieutenants, three engineers, four seaman, two stokers, one carpenter and one steward. Three hands remained aboard *Terra Nova* at Lyttelton as ship keepers. On the whole, for that part of New Zealand, the winter was unfavourable, but in spite of this a satisfactory amount of work was carried out.

On 17 August 1912, while carrying out this work, Leading Stoker Robert Brissenden died. Lieutenant Pennell was not present but the circumstances describing his death are mentioned briefly by Pennell, in an entry in his diary made at Lyttelton, dated 1 November 1912:

> On August 17th, we had the great misfortune to lose Brissenden by drowning. He fell off the pier head and was stunned when he reached the water, but except for this, little light was thrown on it by the Inquest. He was buried on the hillside overlooking the bay and a marble cross has been put up by the expedition. Brissenden was a first class man, careful and reliable, and his loss will be much felt.

News of Brissenden's death was carried to his wife in the UK by Kathleen Scott, at that time unaware that she herself was already a widow. In a letter to her husband, Kathleen Scott wrote:

> Con dear, it was just terrible. She was a dear, quiet little woman with a fine boy

The cross over the grave of Leading Stoker Robert Brissenden. (Marlborough Museum, New Zealand)

of seven. I don't suppose that anybody could have felt her blow more poignantly than I, but I doubt if my sympathy was any help. It seemed so terrible to descend on her gay, neat little house, to drop a bombshell like that, and depart. I wrote to her when I got home, but it all seems so paltry in the face of her sorrow.

The Scott Fund had been set up for those who died on the expedition, and a payment from this was made to the widow of Leading Stoker Robert Brissenden.

# 10

# TO ANTARCTICA – THE FINAL RELIEF

## Last Passage to Antarctica (14 December 1912–12 February 1913)

Terra Nova left Lyttelton at 5 a.m. on 14 December 1912. A crowd of friends had collected to bid the ship farewell. Command of the ship now returned to 'Teddy' Evans, who had returned to New Zealand fully recovered from the effects of scurvy. In his story, *South with Scott* he records, 'She was once more under my command as her original captain, Pennell very gracefully and unselfishly standing down to the position of second in command.'

After leaving the expedition and returning to New Zealand Evans had gone home to recuperate. While in Britain, he had been invited to meet King George V, to whom he related his experiences, including seeing the polar party of five men strike out towards their goal. It was during this return to Britain that Evans was promoted to the rank of Commander RN.

On the evening of *Terra Nova*'s departure from Lyttelton they discovered a stowaway concealed in a lifeboat. On being questioned, he said he was a rabbiter and anxious to make a voyage in *Terra Nova*. He appeared to be about 35 years of age and not very intelligent. They made for the nearest port, Akaroa, to hand him in but met the Norwegian barque, *Triton* at midnight and her courteous captain relieved them of the stowaway, promising to land him at Dunedin.

As they made their way south on the meridian of 165° W, soundings were made to 3,000 fathoms and Lillie made successful plankton trawls. On 17 December, they passed the Antipodes Islands, experiencing heavy seas.

During this voyage, they found the ship infested with rats. Cheetham, the boatswain, who had crossed the Antarctic Circle fourteen times, showed himself adept at rat-catching and soon freed the ship from the pests. He threw rats over the side and the albatrosses and mollymawks would swoop down and devour the vermin in an incredibly short time. All kinds of rat traps were in use and they used mouse-traps to catch the young.

On 26 December, they passed the first icebergs. The great belt of Antarctic ice was reached on 29 December 1912. *Terra Nova* was on the same course as that taken by Sir James Clark Ross, who recorded finding a line of compact hummocky ice in the same position in 1842. The ship's company had expected on this occasion to look forward to an almost ice-free voyage to the Ross Sea, but their hopes were frustrated.

The day after entering the pack they encountered heavy resistance and could only make one knot an hour. They had a tremendous struggle to get into the Ross Sea and had to fight their way for 400 miles to get through, using 7 tons of coal on each daily run. When they collided with a floe, *Terra Nova* shook fore and aft, the officer in the crow's nest experiencing the most violent concussions.

Pennell recorded:

One day a penguin chased us for over an hour, crying out ludicrously whenever one of us imitated its call. The little creature became quite exhausted as it swam steadily after the ship. The poor bird was unable to reach the ship, as the 'kick' of the propeller swirled it away whenever it caught up. The penguin would struggle onto a floe and reel about like a drunken man, until finally it lay still, thoroughly defeated.

The weather by this time was good, which allowed for much scientific work to be carried out on board. Crab-eater seals swimming round the ship were a source of entertainment. Three Emperor penguins were seen on a low iceberg. Two looked very thin and a third, larger one was moulting. The iceberg was very soiled, the penguins must have been there for some time.

Progress was very slow until 16 January 1913, when they cleared the pack only to find themselves in thick fog. They rounded Cape Bird on 18 January and the familiar features of McMurdo Sound were clearly outlined to the south as they steamed at full speed past Cape Royds.

## Cape Evans and the Tragic News

Preparations were made aboard to receive members of the expedition. Mails were sorted and done up in pillowcases, each labelled, the only one piece of bad news being the death of Brissenden. Every telescope and binocular was levelled on the hut at Cape Evans. The view opened up as they rounded Cape Barne. The bay was free from ice and figures were seen outside the hut.

The shore party gave three hearty cheers, to which the ship's company replied. The commanding officer, seeing Campbell, shouted through a megaphone, 'Are you all well, Campbell?'

At this, the men on shore became speechless and after a very marked hush, which quite dampened spirits, Campbell replied:

> The southern party reached the pole on January 18th last year, but were all lost on the return journey, we have their records.

The anchor was dropped, Campbell and Atkinson came aboard and told the story. The cabins that had been prepared for the lost members of the expedition were unmade, flags were hauled down from the mastheads and undelivered mail was sealed for relatives. It was unexpectedly tragic news for *Terra Nova*'s crew.

*Terra Nova* remained off Cape Evans for thirty hours. Preparations were made for the homeward journey. All hands packed and transported the specimens, collections and equipment to the ship. They cleaned the hut and left enough equipment and stores at Cape Evans to see a dozen resourceful men through one summer and winter at least.

This proved to be an act of providence for, only three years later, Shackleton's Ross Sea party of the Imperial Trans-Antarctic Expedition, 1914–17, landed at Cape Evans from the *Aurora* to lay depots to the Beardmore Glacier. This would enable Shackleton to use these supplies on his proposed journey north to McMurdo Sound via the South Pole, having crossed from the Weddell Sea. As previously mentioned, ten members of the Ross Sea party found themselves abandoned at Cape Evans in May 1915, as their expedition ship with eighteen men aboard and damaged steering gear was blown out of McMurdo Sound in a gale. They drifted for ten months through the pack ice and into the Southern Ocean until the crew were able to effect repairs and reach New Zealand.

*Aurora*'s severed anchors remain on the beach at Cape Evans to this day. Had it not been for those deposited stores and equipment supplemented by food and stores left at Cape Royds, the men at Cape Evans might not have survived,

The cross on Observation Hill, McMurdo Sound, looking south toward White Island and Black Island & Mount Discovery. (Michael Tarver, 2000)

for they were not relieved until January 1917. The men had bravely carried out their instructions. The bases toward the Beardmore Glacier were laid as ordered but were destined never to be used.

One man died while laying the bases and later, another two died in McMurdo Sound, before they were eventually relieved. Shackleton, having recovered his men following the loss of *Endurance* in the Weddell Sea, eventually made it to New Zealand and was aboard the refitted *Aurora* when the surviving men of his Ross Sea party were relieved in January 1917. The *Aurora*, with the relief expedition, was under the command of Captain John King Davis.

Shackleton failed to cross Antarctica. His ship, *Endurance*, was crushed by ice in the Weddell Sea. That expedition, in its aims, was a total failure. The story of the loss of *Endurance* and of the Ross Sea party with the loss of three men, together with the drift of the *Aurora* is told by Shackleton himself in *South!*

On 19 January 1913, the expedition aboard *Terra Nova* collected a depot of specimens left by Priestley's party after the Mount Erebus journey from Cape

Royds. The ship then steamed to Hut Point. On 20 January Atkinson, with a party of seven, set out to erect a cross in memory of Scott, Wilson, Oates, Bowers and Edgar Evans. The cross was made in Australian Jarrah wood by Francis Davies, the leading shipwright.

The cross, which is 9ft high, stands on the 900ft summit of Observation Hill overlooking McMurdo Sound. It took two days for the party to carry the heavy wooden cross to the top of the hill. It stands there to this day and the author had the privilege to climb the hill in January 2000 to see the cross, which is inscribed with the names of the polar party and the words, 'To strive, to seek, to find, and not to yield'. The author was able to empathise with those events of ninety years before.

Atkinson and his party returned to the ship, having closed the old *Discovery* hut at Hut Point, where they left a quantity of provisions and equipment. In the meantime, the engineers aboard *Terra Nova* had drawn the ship's fires to clean out the boilers. There was some anxiety as, during the night of the 19th, a large iceberg swept into McMurdo Sound and was carried by the current directly for the ship. Having no steam, they had to set sails. There was some worry, but the sails were just full enough to give steerage way north, allowing the iceberg to pass close across the stern of the ship.

At 5 a.m. on 22 January, the ship proceeded towards Granite Harbour under steam where they collected a geological deposit left by Griffith-Taylor and Debenham. They eventually made their way out of ice fields and towards Terra Nova Bay, near where Campbell and his party had spent the previous winter. A party under Priestley went ashore and collected more geological specimens, while others visited the ice cave where Campbell and his men had lived. Cherry-Garrard's account reads:

> The visit to the igloo revealed in itself a story of hardship that brought home to us what Campbell never would have told. There was only one place in this smoke-begrimed cavern where a short man could stand upright. In odd corners were discarded clothes saturated with blubber and absolutely black. The weight of these garments was extraordinary and we experienced strange sensations as we examined the cheerless hole that had been the only home of six of our hardiest men.

## The Last Departure for New Zealand and Goodbye to Antarctica

Very early on 26 January 1913, *Terra Nova* left the Antarctic continent, never to return. 'Teddy' Evans recorded in his story *South with Scott:*

> We left these inhospitable coasts and those who were on deck watched the familiar rocky, snow-capped shores fast disappearing from view. We had been happy there before disaster overtook our Expedition, but now we were glad to leave and some of us must have realised that these ice-girt rocks and mountains were not meant for human beings to associate their lives with. For centuries, perhaps for all time, no other human being will set foot upon the Beardmore and it is doubtful if ever the great inland plateau will be re-visited, except perhaps by aeroplane. When we left it was a 'goodnight' scene for most of us. The great white plateau and peaks were grimly awaiting winter, and they seemed to mock our departing exploring ship as though glad to be left in their loneland Silence.

When Evans wrote those words, little did he realise the contribution made by the men of those early expeditions. It would lead to an era known as the heroic age of Antarctic exploration and would be the prelude for the programme of science now being conducted by countries which are parties to the Antarctic Treaty. Now, a century later, expedition huts from the heroic age still stand at McMurdo Sound and Cape Adare, each presenting an evocative and haunting scene. They are cared for as listed monuments by the New Zealand Antarctic Heritage Trust, supported by the UK Antarctic Heritage Trust.

As the expedition left the Ross Sea for the last time and made their way north, *Terra Nova* stood well to the westward of the Balleny Islands. On 28 January, Tryggve Gran recorded in his diary:

> Just like a foggy day in the Channel. The sea is running quite high and most of the land party are seasick. The rigging is coated white with ice and rime, and the sails so stiff they are almost impossible to manage. Nelson is on the sick list; his feet are frightfully swollen. I think it's very fortunate we didn't have to face another winter. It's tremendously exciting to be on the bridge. Icebergs loom up out of the murk without warning and many times we've had to spin the wheel hard a' starboard to avoid collision.

As the days went on, he recorded southerly gales. The crew were unable to hoist the sails because of the thick weather and the condition of the ship, which he described as being completely iced up. On 30 January, he recorded a horrible night. The rolling of the ship was so bad he could hardly lie in his bunk. The dogs were extremely disturbed and frightened by the pitching. His dog, Lappa had tried to jump overboard several times.

On 1 February, Gran recorded that the number of icebergs had risen sharply and they sailed alongside one that was 40km long. On 2 February he recorded:

> Poor visibility all day and difficult to navigate. At 11.30 am we entered a swarm of bergs. For a time we were completely surrounded and we only just managed to escape being crushed. I had just relieved Rennick when the look-out shouted, 'Ice ahead' and before I could say anything we appeared to have entered an enormous dock. In my distress I bellowed for 'Teddy' Evans and rang down 'Full astern'. It was an exciting moment. 'Teddy' arrived on the bridge and I must say I was enormously impressed by the way he reacted. His orders flowed as though we were going through a daily drill, though he knew perfectly well he held the fate of 50 men in his hands. We crept along the walls of the 'dock' to find a way out. Luck was with us and we found the right passage at the last moment. We dubbed it, 'Hell's Gate' because we've seen no ice at all since.

As the ship made its way north, Petty Officer Thomas Williamson found time to write a letter dated 1 February 1913, to a lady friend, Miss Lucas. Williamson had been a member of the *Discovery* expedition and joined the shore party of the *Terra Nova* expedition for the second winter. He was a member of the ten-man search party which found the tent of Scott, Wilson and Bowers on 12 November 1912. He wrote:

> We are now on our way to New Zealand and should shortly be on our way home altogether and I can assure you that I shall not be sorry. We have had a bad year of misfortunes and which has been disastrous, as you will see when you read your papers. I do not intend going into details over our sad loss as it is positively painful to me to have to relate sufficient for me to say that our brave fellows and comrades died a most noble death equal to anything that has ever been done on the battlefield, poor fellows how they did suffer, I was one of the few that went in search of them, but only succeeded in finding their frozen remains and their records.

On a port tack and heading for home. (Scott Polar Research Institute)

On 3 February, Gran recorded that they were travelling northwards under full sail and with a bit of luck should reach Lyttelton on Monday. On 4, 5 and 6 February, they found themselves driven east by a strong northerly wind which rose to hurricane force the whole night and through the day on the 5th:

> The dogs were miserable and the wind so strong the men couldn't stand on deck. The sea got up awfully and the ship was tossed about like a cork. Things worked loose and below decks it was chaotic.

Nevertheless, there was time for other things, and in a letter dated 6 February 1913, 'Teddy' Evans wrote to friends Ralph, Sylvia and Lal Gifford, commenting on some aspects of the expedition. He was particularly uncomplimentary about the slow progress of the final polar party on their return journey and the carrying of geological specimens which he felt should have been dumped as 'we dumped all we could'. He referred to the specimens as 150lb of 'trash', all of which could have been recovered later.

To some extent, his letter does raise a question and suggests the somewhat detached part that Evans now took in the expedition. He had been greatly involved from the start. He spent the first winter with the expedition and had been a member of the penultimate polar party. Following his illness and return to Hut Point, bravely brought about by Lashly and Crean, he was fortunate enough to be just in time to catch the return of *Terra Nova* to New Zealand. These events had taken him rather out of the picture.

After his return to New Zealand, he went home to Britain to recover from the effects of scurvy. He returned to New Zealand for the final relief of the expedition and had been promoted from lieutenant to the rank of commander. He had now resumed command of *Terra Nova* and, as Scott's deputy, assumed command of the expedition. He was a fortunate man in many respects and had been spared much of the sad drama of those events of the second winter.

His exaggerated reference to 150lb of 'trash' when, in fact, the weight of the specimens found on the dead polar party's sledge was 35lb, might be considered a personal view, as it was known that his opinion on scientific matters did not take priority. It was subsequently established that these rock specimens brought down from the Beardmore Glacier were Olive Block fossils from cliffs of Beacon sandstone in which Wilson had spotted several plant impressions. Decades later, scientists identified the species as common to other southern hemisphere land masses implying that Antarctica had not always been ice-bound, and supporting the theory of the existence of Gondwana and of continental drift. The value of this find advanced science dramatically.

On 7 February, Gran recorded 'fine weather with a headwind, it was beginning to get warm and nice and they were sunning themselves on deck'. On the 9th February, Gran recorded that they could smell the land and said how wonderful it was to see green fields.

They put in at Oamaru on 10 February, a small port on the east coast of South Island, where Pennell and Atkinson were landed. They carried the official dispatch for *Central News* to distribute the tragic news of the expedition throughout the world.

Cherry-Garrard describes their arrival at 2.30 a.m.:

We crept like a phantom ship into the little harbour of Oamaru on the east coast of New Zealand. With what mixed feelings we smelt the old familiar woods and grassy slopes and saw the shadowy outlines of human homes. With untiring persistence the little light house blinked out the message, What ship's that? What ship's that? They were obviously puzzled and disturbed at getting no answer. A boat was lowered, Pennell and Atkinson were rowed ashore and landed. The seamen had strict orders to answer no questions. After a brief time the boat returned and Crean announced, 'We was chased sorr', but they got nothing out of us. We put out to sea.'

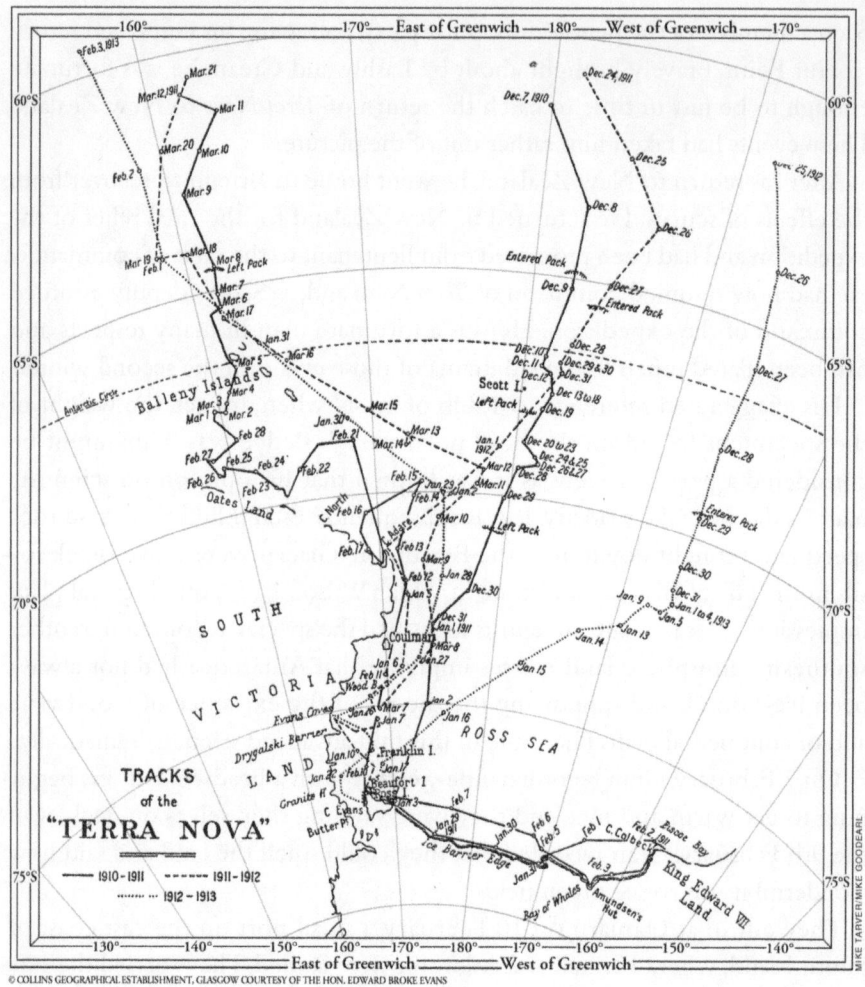

The ship arrived at Lyttelton on 12 February 1913 with flags at half-mast. *Terra Nova* was berthed alongside the Harbour Board shed by Captain Thorpe, the harbour master. Thousands turned out to meet the ship to express their sympathy. Messages of condolences were received from all over the world.

The people of New Zealand marked their association with Captain Scott's last expedition in many ways. At Lyttelton Harbour, a bronze plaque is mounted at the berth where *Terra Nova* and the expedition made their first call in New Zealand to take on supplies and prepare equipment and there are other commemorative plaques in the town. The dry dock where work on *Terra Nova* was carried out still exists. A statue of Captain Scott by

Kathleen, his widow, stands on the banks of the River Avon in the nearby city of Christchurch.

At Port Chalmers, a memorial cairn was formally unveiled on the hill overlooking the harbour on 30 May 1914. In the suburb of Roslyn in Dunedin, Falcon Street and Oates Street are named in memory of the explorers, and the Port Chalmers Monowai Sea Scout Group took the name *Terra Nova* in their title.

Though the call at Oamaru was a brief stop only to deliver the news agency report, the port proudly recorded the association with the expedition by planting a memorial oak tree in Arun Street. The Sea Scout Group there took the name 'Scott's Own' in its title.

# 11

# THE VOYAGE HOME

## The Passage to Britain (13 March–14 June 1913)

*Terra Nova* sailed from Lyttelton for her homeward voyage to Cardiff on 13 March 1913, once more under the command of Lieutenant Pennell. In the wardroom were the captain, Henry Rennick, Edward Nelson, Dennis Lillie, Murray Levick, Gibson Anderson, William Williams and Alfred Cheetham. Edward Nelson volunteered as second mate and Gibson Anderson of Christchurch was taken on for coal trimming. The ship had thirteen dogs aboard, many as pets of its members. Special platforms were built for them on deck, giving comfort in both hot and poor weather.

The journey across the South Pacific began with fog and thick weather and they were driven further south than was intended. Once more they saw icebergs. On 29 March they were able to alter course north-east and leave behind those uncomfortable latitudes.

Cape Horn was passed in a strong gale on 11 April, but thereafter, with little exception, fine weather was

The Scott Memorial Cairn at Port Chalmers, unveiled on 30 May 1914. (Courtesy Port Chalmers Museum)

experienced all the way home. Lillie carried out some trawling, and much interest was created aboard with everyone crowding round to see what specimens had been dragged up from the bottom of the sea.

They stopped at Rio de Janeiro on 28 April for four days to take on coal and provisions, and for leave. They called at the Azores in order to cable home.

On 11 June 1913, *Terra Nova* dropped anchor in Crow Sound, Scilly Islands, where two days were spent painting and cleaning the ship. Here, Lieutenant Pennell handed over command to 'Teddy' Evans for the passage up the Bristol Channel to Cardiff.

Lieutenant Pennell finally recorded:

Here, it only remains to acknowledge the exemplary conduct of the ship's company, fore and aft. Every member worked to help the Expedition forward loyally and cheerfully, accepting each position as it came, all hands doing their best to help matters forward and to see the humorous side of everything.

## Return to Expedition Home Port

On Thursday, 12 June, *Terra Nova* was boarded at Crow Sound, Scilly Islands, by Commander Evans for the passage up the Bristol Channel to Cardiff. Awaiting the return of the expedition were relatives and friends, some of whom were ferried out in the tugs *Nelson* and *Mercedes* for the reunion. Many thousands of spectators were on shore.

*Terra Nova* arrived at Cardiff on 14 June 1913 and entered the docks through the same lock gates of the Roath Basin from where she had departed almost three years before to the day. She flew from the foremast the city coat of arms and the Welsh dragon on the main mast. Scouts of the 4th Cardiff (St Andrews) Group provided a guard of honour. Accompanied by a goat mascot, the Scouts lined the quayside to await the return of their colour, which had been taken south aboard the ship.

It was a sombre return to Cardiff compared with the rapturous departure that the expedition had enjoyed three years before. The Welsh dragon was given to the National Museum of Wales, the city flag was given to the city authorities and the White Ensign was offered to Daniel Radcliffe as recognition of his contribution to the expedition. The burgee of the Royal Yacht Squadron is now at the Captain Oates Museum at Gilbert White's House, Selborne, Hampshire.

Scenes aboard *Terra Nova* at Lyttelton on return from the south. Mr Joseph Kinsey is to the right of Commander Evans. (Courtesy Hon. Edward Broke Evans)

Peter Scott, the son of Captain Scott, was 10 months old in 1910 when his father's expedition left for the Antarctic. Sir Peter recalled, years later, to Dr Anthony Johnson, author of *Scott of the Antarctic and Cardiff*, his memory of the ship's return to Cardiff in 1913 and of being carried up to the crow's nest of the *Terra Nova* by Tom Crean.

Here at Cardiff, the expedition was to be paid off. A reception was held aboard for invited guests and a recollection of that event by the late Mr George Buckmaster was often proudly recounted to his daughter, Mrs June Siveyer Watson, now of Ashburton, south Devon. At that time, George Buckmaster was manager of Wyman & Sons Ltd, newsagents, publishers and printers, at their Cardiff branch, 56 St Mary Street, in the city centre. Together with a photographer, he had gone aboard *Terra Nova* and presented a substantial donation on behalf of his company to be given either towards the expedition's outstanding funds or as a contribution to the National Appeal Fund launched for the dependants of those who had died. Consequently, it was arranged by Mr Daniel Radcliffe for Mr Buckmaster to be loaned items for display at his shop, including sleeping bags, sledging equipment and many other effects used by the expedition, all very much to the interest of customers and passers-by who saw the display in his shop window.

*Terra Nova* at Crow Sound, Isles of Scilly, 12 June 1913. (Courtesy of Sandra Kyne, St Mary's, Isles of Scilly)

Roath Basin, Bute Docks, Cardiff, as *Terra Nova* returned on 14 June 1913. (Courtesy *South Wales Daily News*)

After its return, this colour was placed in a display cabinet at the Scout group's headquarters and has been a treasured item in their archives ever since. It even survived an air raid. On 30 April 1941, during a night raid on Cardiff by the Luftwaffe, the group's headquarters building at Wyverne Road was destroyed by a land mine. Amazingly, the colour in its display case survived undamaged and to this day it is still on display at its present headquarters. Such good fortune did not repeat itself later, though. When on 10 November 1984, the 'In the Footsteps of Scott' Expedition, led by Robert Swan left Cardiff in the ship *Southern Quest* as part of the re-enactment, this Scout group again presented a colour to the expedition, but it was never to return. Robert Swan's expedition reached the South Pole, but *Southern Quest*, awaiting their return to base, was crushed by ice and sank near Beaufort Island in the Ross Sea, just north of McMurdo Sound. The colour went down with the ship. (4th Cardiff Scout Group archives)

When the ship was unloaded, expedition equipment was shared between the crew members as souvenirs and much was given away to local Scout groups and other organisations. Many years later, Scouts still made use of Antarctic pyramid tents, which were pitched in a churchyard north of Merthyr Tydfil. A pair of skis that are said to have been used by Tom Crean can be seen at the Cyfartha Castle Museum & Art Gallery, Merthyr Tydfil.

On 16 June 1913, a dinner was held at the Royal Hotel, Cardiff, by the commercial community of the city 'To honour the Officers and Crew of the 'Terra Nova' and the British Antarctic Expedition, 1910–1913'.

While the ship was open to members of the public, the *Western Mail*, in the issue dated 7 July 1913, reported that there had been a surprise find aboard the *Terra Nova*. Someone had accidentally disturbed a loose wooden

Display of expedition equipment at Wyman's Newsagents, 56 St Mary Street, Cardiff. (Courtesy June Siveyer Watson)

Invited guests boarding *Terra Nova* on her return to Cardiff. (Courtesy June Siveyer Watson)

board in the cabin used by Captain Scott and had discovered a considerable quantity of tinned provisions!

The ship was prepared for her departure from Cardiff. Ownership of *Terra Nova* was to be returned to C.T. Bowring & Co. Ltd, as arranged under the purchasing agreement.

A number of events subsequent to the expedition's association with the city of Cardiff took place later in 1913 and in following years. The figurehead was taken from the bow of the ship and presented as a gift to the city and port of Cardiff by C.T. Bowring and Co. Months later, in a formal ceremony at Roath Park, Cardiff, on 8 December 1913, the figurehead and an inscribed plaque were unveiled by Mr Frederick Charles Bowring. At that ceremony Mr Bowring suggested that *Terra Nova*, because of her ownership and her Welsh commander, might be looked upon as a Cardiff vessel. He announced that he would erect, at his own expense, a monument in the form of a clocktower to serve as a permanent memorial to Scott and his companions and to record the connection of the expedition with the city of Cardiff.

The clock tower memorial lighthouse, with a bronze plaque showing the names of the five men who died in the Antarctic, was unveiled by Mr Bowring at a formal ceremony on

*Above: Terra Nova*'s figurehead. (Courtesy *Western Mail*)

Sir Frederick Charles Bowring DL JP (1857–1936), eldest son of John Bowring (1824–86) who was the fifth son of the company founder, Benjamin Bowring. (Courtesy of the Bowring family)

*Above:* A reception aboard *Terra Nova*, 1 July 1913. Front row: Mr F.C. Bowring, Bosun Cheetham, Petty Officers Forde & Bailey, Commander Evans, Mr Daniel Radcliffe. Back row: Petty Officers Lashly, Keohane and Crean. (Courtesy June Siveyer Watson)

*Left:* Petty Officers William Lashly & Thomas Crean, who saved the life of Lieutenant Evans, aboard *Terra Nova* on 14 June 1913. (Courtesy National Museum of Wales)

14 October 1918. It stands to this day in the beautiful setting on the promenade at the southern end of Roath Park Lake, Cardiff. A model of a ship representing *Terra Nova* serves as a wind vane. A tablet in honour of Captain Scott stands on the stairway of the City Hall.

At the end of the First World War some residents of Cardiff and friends who had known expedition members presented plaques and endowed beds at the Royal Hamadryad, a seaman's hospital at Cardiff docks, in honour of expedition members. These were in honour of Lieutenant Henry E. de P. Rennick, lost in the cruiser HMS *Hogue*, torpedoed and sunk with three other British warships in an engagement off the Dutch coast on 22 September 1914 and Commander Harry L.L. Pennell, navigating officer of the cruiser HMS *Queen Mary*, sunk by enemy gunfire at the Battle of Jutland on 31 May 1916, as well as the brave deeds of Commander E.R.G.R. Evans for his wartime exploits as captain of HMS *Broke* in 1917.

After naval service in the First World War, two members of the expedition settled in Cardiff and worked at Cardiff docks. Petty Officer William Lashly was aboard HMS *Irresistible* and survived when she was sunk during the Dardanelles campaign. After further service in the Mediterranean in HMS *Amethyst*, Lashly returned to Cardiff at the end of the war and took up an appointment as engineer with the Board of Trade. He lived at 17 Mayfield Avenue, Heath, Cardiff, until he returned to his home village of Hambledon, Hampshire in 1932.

James Skelton AB served in the Royal Naval Reserve on armed trawlers in the North Sea and returned to Cardiff after the war. He married and lived at 32 Stockland Street, Grangetown, Cardiff, taking up employment at the docks on local ships and then in the marine protective clothing trade.

After many years in the open air at Roath Park, the figurehead of *Terra Nova* passed into the possession of the National Museum of Wales in Cardiff. The ship's bell, which was acquired by Surgeon Lieutenant Atkinson, was later presented to the Scott Polar Research Institute, University of Cambridge, where it is rung twice daily to announce the staff tea break, five bells at 10.30 a.m. in the Forenoon Watch and eight bells at 4 p.m. in the Afternoon Watch.

## Summary of British Antarctic Expedition Programme, 1910–13

It is not intended to include a full account of this famous expedition in the story of the *Terra Nova*, but a summary of principal events which occurred during the expedition is set out below to link with the voyages of the ship.

The shore party at Cape Evans, as well as establishing their base, began preparations for the depot-laying journey almost immediately on arrival. With the departure of *Terra Nova* with two scientific parties aboard to be disembarked at given locations, the party led by Scott began laying depots south. It was intended to reach Lat. 80 ° S and establish a substantial base as a preliminary to the main polar journey which was to begin later that year, in November, 1911.

This party consisted of twelve men, eight ponies and twenty-six dogs. In the event, the depot was laid at Lat. 79° 29' S, 31 miles short of the intended Lat. 80°. Scott wrote in his journal, 'It is a pity we couldn't get to 80 degrees … The party had an eventful return journey, with narrow escapes for the men and with the loss of some ponies and dogs.'

Preparations were made at Cape Evans for the first winter and it was during this period that Wilson, Bowers and Cherry-Garrard made the winter journey to Cape Crozier to study the Emperor penguin. This was often described as the 'weirdest bird-nesting expedition' ever. They returned safely to Cape Evans after a six week journey there and back in an Antarctic winter and were able to relate an incredible account of hardship and endurance, but the journey was scientifically successful.

The sixteen-man southern polar journey commenced in stages. The two motor sledges departed in the charge of four men on 24 October 1911. Ten men and ten ponies departed on 1 November while two men ran teams comprising thirty-two dogs. From Hut Point to the South Pole and back is 1,532 geographical miles or 1,766 statute miles.

The motor sledges were abandoned after the first 50 miles, and later two men from the motor party returned to Cape Evans. The use of all animals was dispensed with at the foot of the Beardmore Glacier, where the last of the ponies were shot. The dogs and their two handlers returned to Cape Evans.

On 9 December 1911, twelve men began the ascent of the Beardmore Glacier and on reaching the polar plateau, eight men continued south, leaving another four men to return. On 4 January 1912, 150 miles from the South Pole, Scott selected four men to accompany him south, sending three men back. He chose Wilson, Oates, Bowers and PO Evans to accompany him; Lieutenant Evans, Lashly and Crean were sent back.

On 16 January 1912, the five man polar party came across the black marker flag of Roald Amundsen, indicating the presence of the Norwegians. On 17 January 1912, Scott and his party arrived at the South Pole to discover the Norwegian tent and its contents, and to suffer the disappointment of not gaining the honour of being first to the South Pole. After spending one night at the pole, Scott and his party turned to begin the return journey.

Meanwhile, on 4 February 1912, *Terra Nova* returned to Cape Evans from New Zealand.

On 22 February 1912, the penultimate polar party of three men – Lieutenant Evans, William Lashly and Tom Crean – had returned to base, but not without drama. They struggled on their homeward journey and only the heroic efforts

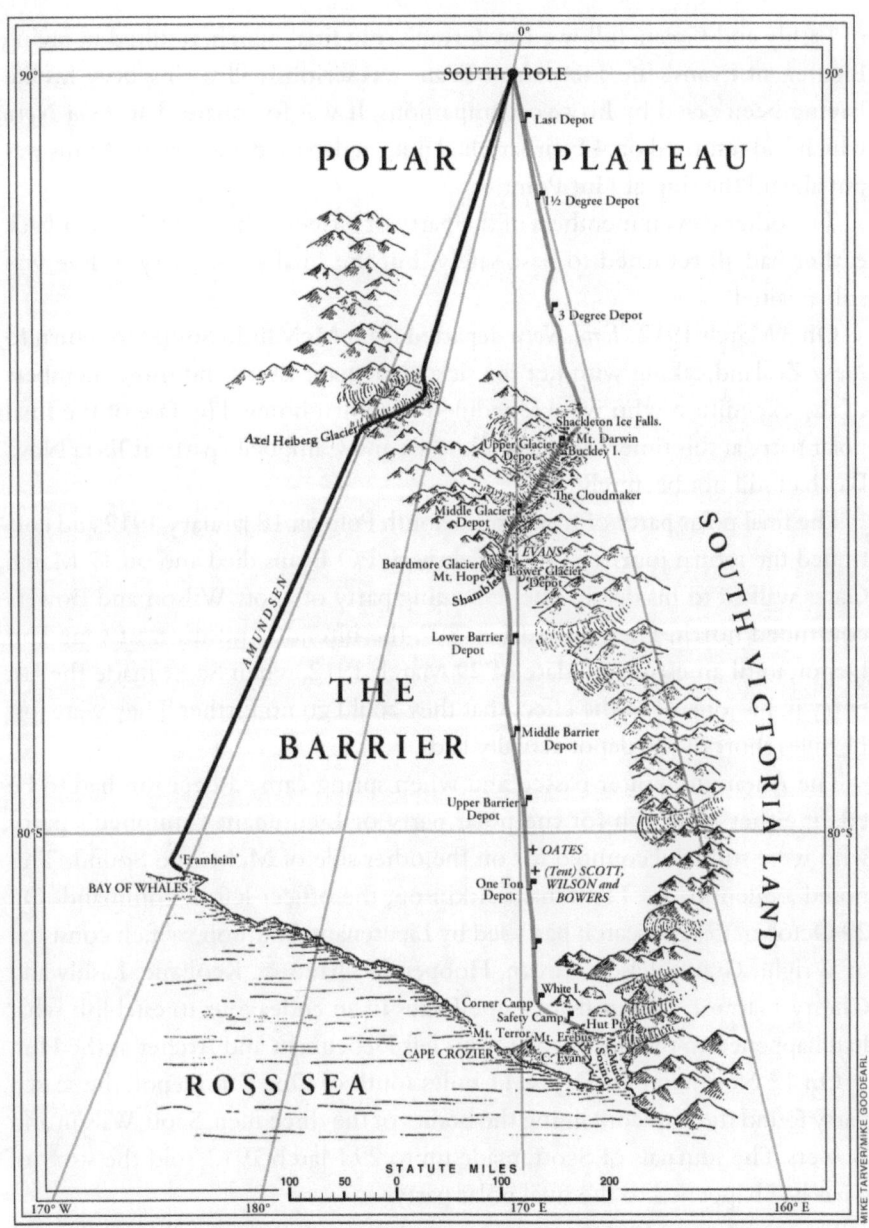

Routes to the South Pole taken by Amundsen and Scott, 1911–12.

of Lashly and Crean, followed by Crean's solo final march, resulted in saving Lieutenant Evans's life. Lieutenant Evans was seriously ill with scurvy, his life having been saved by his two companions. It was fortunate that *Terra Nova*, which had returned on 4 February, had not yet departed. Lieutenant Evans was put aboard the ship at Hut Point.

The other eleven members of the party of sixteen who had been sent back earlier had all returned to base safely, but the final polar party of five was still awaited.

On 3 March 1912, *Terra Nova* departed from McMurdo Sound to return to New Zealand, taking with her the sick Lieutenant Evans and those members of the expedition who were scheduled to return home. The fate of the final polar party at this time was still not known and Campbell's party at Terra Nova Bay had still not been relieved.

The final polar party of five left the South Pole on 18 January 1912 and continued the return journey. On 17 February, PO Evans died and on 17 March, Oates walked to his death. The remaining party of Scott, Wilson and Bowers continued north. They passed Lat. 80°, the intended position of One Ton Depot, until an estimated date of 29 March 1912, when Scott made the last entry in his journal to the effect that they could go no further. They were just 11 miles short of the depot actually laid.

The Antarctic winter passed and when spring came a decision had to be taken either to search for the polar party or Lieutenant Campbell's party, who were still unaccounted for on the other side of McMurdo Sound. This posed a dilemma for Lieutenant Atkinson, the officer left in command. On 29 October 1912, a search party led by Lieutenant Atkinson, which consisted of Wright, Gran, Nelson, Crean, Hooper, Williamson, Keohane, Lashly and Cherry-Garrard, set out from Cape Evans in an endeavour to establish what had happened to the polar party. This left Debenham and Archer at the Hut.

On 12 November 1912, just 11 miles south of One Ton Depot, the search party found the tent containing the bodies of the three men, Scott, Wilson and Bowers. The journals of Scott, made up to 29 March 1912, told the story of what had happened to the final polar party.

The search party took possession of personal articles and collapsed the tent over the three bodies. A cairn was built surmounted with a cross and a short service was held in honour of the dead explorers. When they returned to Cape Evans, the search party was relieved to find that Campbell and his party had been able to make their own way back from the other side of McMurdo Sound, where earlier in the year *Terra Nova* had found it impossible to relieve them, in spite of a number of attempts which were foiled by sea ice.

On 18 January 1913, *Terra Nova* returned to Cape Evans from New Zealand to make her final relief of the expedition and bring them home. The crew then learned of the tragic polar journey. Commander Evans, who had now recovered from scurvy, took command of the expedition.

A party from the ship climbed Observation Hill on 20 January and erected a cross in memory of the dead polar party.

Finally, on 26 January 1913, *Terra Nova* departed from McMurdo Sound and returned to Lyttelton, New Zealand, where the tragic news of the expedition was made known to the world.

On 13 March 1913, *Terra Nova* departed from New Zealand under the command of Lieutenant Pennell and returned to Cardiff, where she arrived on 14 June 1913. Here the expedition paid off.

In a ceremony at Buckingham Palace on the 26th July 1913, members of the expedition paraded before King George V. The king presented the Albert Medal* to Petty Officers William Lashly and Tom Crean for saving the life of Commander Evans. Of the sixty-five-man expedition, Bronze Polar Medals were awarded to the six men who had made one passage to Antarctica and the Silver Polar Medal was awarded to those who had made two journeys south. A Silver Clasp was awarded to those who already held the Silver Polar Medal.

On 15 August 1913, *Terra Nova* left Cardiff for Newfoundland to resume her role in the sealing industry.

# Report on the Biological Work Carried Out Aboard by Dennis G. Lillie, MA

*From the Report Published in* Scott's Last Expedition *by Leonard Huxley (1913)*

Captain Scott, with his characteristic thoroughness, made it possible for scientific work to be undertaken out by the ship's party not only on the three summer voyages to the Antarctic, but also during the two winters spent in New Zealand and during the outward and homeward voyages. As Scott's Last Expedition by Leonard Huxley, pub. 1913, was published so early, it was

---

* In 1971, holders of the Albert Medal were permitted to exchange it for the George Cross, and those who did so were reinvested.

impossible to give any adequate account of the various biological results which might have been achieved. It is proposed to give here a brief summary of the collections brought home, together with a few notes concerning them, to help the general reader to form some idea of what he will find in the results of the Biological Reports of this expedition when they appear.

Biologist Dennis G. Lillie. (Herbert Ponting)

<u>The Outward and Homeward Voyages</u>
Whenever opportunities occurred on the outward and homeward voyages between Britain and New Zealand, tow-nets of fine mesh and of various sizes were put overboard to catch the small animals and plants which drift about in the sea and form the staple food of the baleen whales and of many birds and fishes.

These floating organisms, which include representatives of many of the larger divisions of the animal kingdom, are spoken of collectively as the plankton. On some occasions the net was towed behind the ship for about half an hour to catch the floating population of the surface. Sometimes the ship was kept stationary and the net sent down by a sinker to 500 fathoms or more and hauled up again; by this means samples of those forms which live below the surface were obtained.

About 70 samples of the plankton were collected. They vary greatly in size, one catch hardly covers the bottom of a half-pound honey jar, while another requires two seven-pound fruit jars to contain it.

The size of a catch depends on various factors, such as the size of the net, the time it was fishing or the amount of water passing through it and the quantity of plankton in the sea at the place where the haul was obtained.

A small number of sea-water samples were collected from various depths by means of the Nansen-Pettersson water-bottle. These were generally taken from the areas in which plankton samples were obtained. The object of these water samples is to ascertain the salinity of the sea at different positions and at different depths. Any change in the salinity means a marked change in the character of the plankton.

The plankton catches, when sorted, will doubtless be found to contain many new genera and species to add to the list of the known forms of living things. The vertical hauls, which were generally made for quantitative purposes, will help to increase our knowledge of the relative abundance of the plankton over the oceans of the world and at different seasons of the year. Isolated observations such as these may be of small value in themselves, but every expedition which collects such data thereby adds its quota to the gradually accumulating mass of evidence and brings the time for generalisation nearer to hand. A knowledge of the relative abundance of the food supply of the ocean is not only of scientific interest but of commercial importance.

On the homeward voyage two satisfactory hauls with the trawl were obtained, one off the Falkland Islands in a depth of 124 fathoms and the other off Rio de Janeiro in 40 fathoms. The trawl scrapes the bottom of the sea, and brings up a fair sample of whatever animals and plants it can entrap or uproot.

So little scientific trawling has been done in the Southern Hemisphere that almost every haul has a chance of containing some creature hitherto unknown to science from the area in which the catch was obtained.

Animals which live at the bottom of the sea are known to zoologists as the benthos.

During the outward voyage a day was spent on the island of South Trinidad by several members of the Expedition, and collections of land plants, land spiders, insects, and marine coastal animals were obtained.

The collection of plants has been examined by Dr. O. Stapf of Kew Herbarium, and found to contain some thirteen species which have not hitherto been recorded from the island. South Trinidad is a small volcanic island lying about 500 miles from the coast of Brazil, whence it has derived its scanty fauna and flora by means of such agents as winds, ocean currents and birds. On the voyage home we actually saw one of these agencies at work. When the ship was rather more than a hundred miles from the Brazilian coasts, to the southward of Trinidad, a large number of moths, belonging to about four species, were blown aboard by a S.W. wind.

An up-to-date account of the fauna and flora of this island will be included in the Reports.

## New Zealand

*Terra Nova* was engaged upon her three months' surveying the vicinity of the Three Kings Islands, off the extreme north of New Zealand, some 80 samples of plankton and 32 samples of sea-water were obtained.

Seven successful trawls in depths varying from 15 to 300 fathoms yielded a good collection of benthos from this area.

During the first winter the ship's biologist spent five weeks at Mr. Cook's whaling station near the Bay of Islands in the north of New Zealand; and in the second winter, through the kindness of Mr. L.S. Hasle, four months were spent on two Norwegian floating factories which were exploiting the same waters. Three specimens of baleen whales were examined and found to be identical with the three northern species. *Balaenoptera Sibbaldai*, the Blue Whale; *B.borealis*, Rudophi's Rorqual; and *Megaptera longimana*, the Humpback Whale. About 30 specimens of the last species were examined. An embryo two and a half inches long was obtained from a female humpback whale weighing about 60 tons.

While at the Bay of Islands an opportunity was taken to examine the inheritance of the pigment in several families of Maori–European half-castes. Sufficient data were collected to show the phenomenon of Mendelian segregation evidently takes place.

Collections of fossil plants were made from several localities in the South Island of New Zealand, with a view to settling the geological age of the so-called 'glossopteris beds' of Mt. Potts. From this material Dr. E.A. Newell Arber has been able to show that the oldest known plant-bearings beds in New Zealand are of Rhaeto-Jurassic age.

One volume of the Reports will be devoted to a description of these fossil flora, together with the fossil plants found by the Polar Party and others in the Antarctic.

An account of some undescribed collections of New Zealand Tertiary and Mesozoic marine invertebrates is to be included in the Expedition Reports.

## The Antarctic

During the three summer voyages to the Antarctic a series of qualitative and quantitative plankton samples were taken between New Zealand and McMurdo Sound, also in different parts of the region visited by the ship.

The number of plankton samples obtained was 135. These comprise 27 collected during the first year, 48 the second year, and 60 the third year. Also 96 samples of sea-water were obtained.

The increase in the relative size of the plankton catches as we left the warm seas around New Zealand and entered the cold waters of the far south was very marked. This increase was especially noticeable in the case of the diatoms. These minute protista became so numerous as to choke the meshes of the net after it had been fishing only five minutes. In the middle of pack ice the diatoms were much less numerous. This may have been due to the ice floes shutting out the sunlight or to an alteration in the salinity of the sea caused by the melting of the ice.

Some fifty samples of the muds and oozes from the bottom of the sea between New Zealand and the Antarctic were collected. A rough examination of some of these showed them to consist of the skeletons of diatoms and other inhabitants of the surface water which had fallen to the bottom. These samples were obtained by letting down a weighted tube on the end of the sounding wire. The tube would sink vertically into the mud and bring up several inches of the deposit. Thus, if there were six inches of mud in the tube a sample taken from the bottom of the tube would come from about six inches below the surface of the sea floor.

In the Ross Sea it was found that many of the diatoms in a sample of mud taken from four inches below the surface of the deposit still contained their protoplasm and chlorophyll bodies. In other words, they were undecomposed.

When trawling in McMurdo Sound it was a common occurrence to find that nearly half the catch consisted of dead animals.

From an examination of the summer temperatures at various depths in several parts of the Ross Sea it was found that a temperature of + 1 degree Centigrade was hardly ever reached. The usual temperature was slightly below 0 degree Centigrade.

At these low temperatures bacterial decomposition is minimal and the food supply of the ocean remains cold. However, a small amount of decomposition must take place to allow the production of nitrates for the plants.

The abundance of plankton in Antarctic waters is shown by a brownish discolouration of the sea produced by the diatoms.

Another indication is given by large numbers of baleen whales, which feed upon the plankton.

It is true that only about three species of baleen whales were recognised south of the pack, but the number of individuals seen daily around the ship was very great. The two commonest species seen were *Balaenoptera Sibbaldi*, the Blue Whale, and *Balaenoptera rostrata*, the Pike Whale.

The large schools of killer whales, *Orca gladiator*, are an indirect indication of a plenteous food supply, because they feed upon seals and penguins, which in their turn live upon the plankton.

If it was fully realised by whalers that there is a natural reason for the abundance of whales in the cold waters of the Polar regions, they would not exploit warm seas such as that off the north of New Zealand with a 'trying out' plant suitable for South Georgia or the South Shetlands and so lose large sums of money.

Fifteen rich hauls with trawl and dredge in depths varying from 40 to 300 fathoms enable a large collection of the benthos to be made. A striking feature of the marine fauna of the Antarctic is the extraordinary wealth of individuals, while the variety of forms does not appear to be very great. Also the large size to which some species attain as compared with their relatives in warmer seas is very marked.

This is, however, not the case with animals which require carbonate of lime, for the secretion of limy skeletons by members of the benthos seems to be at a minimum in the cold Antarctic waters. The shells of molluscs are small and fragile. Some sea-snails have no lime in their shells at all.

It requires the warm tropical seas for animals with calcareous skeletons to reach their vigorous growth.

Many of the bottom animals crawl over the sea floor and pass the nutritious mud through their digestive organs after the manner of earth worms; others take up a stationary vertical position, and by means of tentacles waft the falling diatoms into their mouths before they have time to reach the bottom.

Almost every trawl brought up quantities of large siliceous sponges covered with glassy spicules. Good collections of sea-anemones, worms, urchins, star fishes, crustacea, sea-spiders, molluscs, and fishes were obtained. The collection of fishes has already been found to contain some new genera and several new species. There can be no doubt that many new forms will be found among the other groups.

Considerable quantities of three species of *Cephalodiscus* were obtained. These animals are of interest because they show signs of a distant relationship to the vertebrates, though their mode of life is very dissimilar. The minute individuals live together in colonies and built up a gelatinous tree-like house.

The young forms of *Cephalodiscus* are very imperfectly known, and it is hoped that larval stages may be found among the material brought home, so that further light may be thrown upon the development of these curious animals.

In the last volume of the Biological Reports it is proposed to review the known marine benthos of the continental shelves of the globe in regard to its distribution in time and space. One of the objects of this enquiry will be to ascertain, as far as our present knowledge will permit, if there has been any tendency on the part of the benthos to originate in the Northern Hemisphere and migrate southward.

The work of Wallace on the distribution of land animals has shown that there appears to have been a tendency throughout the history of the earth for the land animals to originate in the Northern Hemisphere and gradually find their way south. The great belt of land which through long ages has almost encircled the northern half of the world seems to have been Nature's workshop for the evolution of types. These new forms spreading out from their points of origin had to find their way southward along the attenuated land areas of South America, Africa, and Australasia. Thus on account of their relative isolation these three southern continents became characterised by peculiar and in some cases comparatively primitive assemblages of animals. They became, as it were behind the fashion. For instance, Australia today still has its marsupial population of kangaroos and such-like animals. In Europe marsupial types are only found as fossils showing that they lived here millions of years ago in the Mesozoic ages of the earth's history, but have long since been exterminated and supplanted by newer types.

On account of the inadequate nature of the fauna of large parts of the Southern Hemisphere, man as had to stock these lands with northern animals. Very few cases are known where land animals of a southern origin have advanced northwards. Whether this generalisation applies in the case of the marine benthos of the continental shelves presents an interesting field of inquiry. The collections brought home in recent years by the various Antarctic and other expeditions which have trawled in the Southern Hemisphere will, perhaps, make it possible to give some sort of answer to this question.

(Marine Biology was also carried out by Edward W. Nelson at Winter Quarters, Cape Evans, between 1911 and 1913)

## Extract from Report: 'Outfit and Preparation' by Commander E.R.G.R. Evans

On the arrival of *Terra Nova* at Cape Evans on 18 January 1913 for the final relief, Commander 'Teddy' Evans, as deputy leader, assumed command of the expedition on hearing of the death of Scott.

In *Scott's Last Expedition* by Leonard Huxley, in the chapter entitled 'Outfit and Preparation', Commander Evans in his report summarised the formation of the expedition and its many contributors. He described the acquisition of the *Terra Nova*, its fitting out, modifications, aspects of general funding and of the applications received from 8,000 volunteers to join the expedition.

The last paragraph of this report reads as follows:

Silver tableware from RYS *Terra Nova*. (Michael Tarver, courtesy Hon. Edward Broke Evans)

*Terra Nova*'s figurehead and the memorial clock tower lighthouse at Roath Park Lake, Cardiff, around 1920. (*Western Mail*)

# THE BRITISH EXPEDITION TO THE SOUTH POLE.

## HOW THE "WESTERN MAIL" HELPED CAPTAIN SCOTT.

By Capt. E. R. G. R. EVANS, C.B., D.S.O., R.N.,
Late Commander of the S.S. Terra Nova, and Second in Command of Captain Scott's Expedition,
now Senior Naval Officer at Ostend.

I cannot let the occasion of the Jubilee of the *Western Mail* pass without sending you a birthday message.

The Jubilee of the *Western Mail* reminds me of the great assistance your paper rendered in helping the late Captain Scott to raise funds for his expedition.

It is well known to the *Western Mail* readers that originally a Welsh expedition was being organised and prepared by me, but that, on hearing that my late leader, Captain Scott, was proposing to lead another expedition to the Antarctic, I decided to forego my own expedition and join forces with Captain Scott.

I, therefore, threw all my energies into raising funds for his expedition. The people of South Wales, thanks largely to the efforts of the *Western Mail*, contributed most generously through me to the new expedition. Captain Scott was so grateful for the assistance he received that he decided to make Cardiff the final port of departure of his expeditionary ship, the Terra Nova.

I must remind you of the interview you had with my late leader, when he gave you the impression that my Celtic exuberance was responsible for the assurance that Wales would do great things for his expedition.

You will remember that Captain Scott did not eagerly accept your proposal to allow me to come down to South Wales in order to place the matter before Welshmen. After pondering over your suggestion you will remember that he said, " I don't want Evans to go down to Cardiff for £25 or so." We all know what the consequences of my visit were, what immense support Cardiff gave to the movement, and the association between Cardiff and our expedition is now historical.

One night before the Terra Nova left Cardiff on her long voyage Captain Scott said at a public banquet :—

We could not have faced the strain of preparation except for the support we received from South Wales. When we entered upon the preliminary work of the expedition nine months ago I secured the services of a Welshman (Lieutenant E. R. G. R. Evans), whom I knew well, and I sent him down to Cardiff to canvass. The result of that canvass was that we have been able to fit up a jolly good expedition, and I cannot find words to express adequately my appreciation of Lieutenant Evans's services.

But I, in my turn, must say that the £26,000 raised through Welsh influence could never have been raised by me without the tremendous assistance I received from the editor and staff of the *Western Mail*.

Looking through some of my old papers, I find a copy of a letter sent you by Captain Scott. He says, among other things :—

I should be lacking in gratitude if I did not, even at this late moment, express my appreciation of the particular service which has been rendered to our cause by the *Western Mail*.

Throughout the preparation of the expedition you have freely used your great influence to interest the public in the venture, and to gain its support for our needs.

I most gratefully acknowledge that this great assistance has been given freely from a patriotic desire to advance a national undertaking.

I dare not trespass any more upon your space, but I am glad to have this opportunity of wishing the *Western Mail* continued success.

Scarcely a week of the great war went by without my receiving one or two copies of your paper, containing some information or other that was interesting.

*Edward R. G. R. Evans*

---

It only remains to acknowledge the unbounded hospitality of the Cardiff citizens, with Mr. Dan Radcliffe at their head, who docked and coaled the ship for us, gave freely in money and kind, and made their generosity so felt that Captain Scott promised that Cardiff should be the home port of the *Terra Nova*.

After the First World War, the *Western Mail* published a letter, reproduced here, from Captain E.R.G.R. Evans RN on 2 May 1919, during the twenty-fifth year of the newspaper's publication. In this letter, 'Teddy' Evans wrote in gratitude of the support which the expedition had received from the *Western Mail* and from the people of South Wales.

# 12

# NEWFOUNDLAND AND SEALING

## Thirty Years with the 'Wooden-Walls'

On 15 August 1913, at 6 p.m., *Terra Nova* made a quiet and lonely departure from Cardiff. This time, there were no cheering crowds, just groups of onlookers. At Newfoundland, the ship was destined to resume her role as a sealer and coastal trader. A crew had been sent to man her from St John's by Bowring Brothers Ltd and for her return she was commanded by Captain James Paton* of Rock Ferry, Liverpool.

The *South Wales Daily News* for Saturday, 16 August 1913, reported:

> The Antarctic ship, '*Terra Nova*' left Cardiff yesterday for St John's, Newfoundland to re-engage in the whaling and sealing industry. After a refit, she was found in capital condition showing little or no trace of her stressful time in the Antarctic, though her figurehead was missing from the bow having been presented to the City of Cardiff. She took away a cargo of coal and was under the command of Captain Paton of Liverpool.

She arrived at St John's on 31 August 1913.

---

\* No known connection with the James Paton who previously served on the *Terra Nova* and other vessels on expeditions during the heroic age.

Captain James Paton and crew departing for Newfoundland on 15 August 1913. (*South Wales Daily News*)

A letter dated 16 July 1962, from C.T. Bowring & Co. Ltd, enclosing a memorandum from Bowring Brothers Ltd at Newfoundland, states that *Terra Nova* was square rigged on her return but the yards were later removed. The memorandum also states that some years later a replacement boiler and smokestack were fitted. These were taken from the naval vessel HMS *Lobelia*, once owned by the Newfoundland Government, but subsequent to dismantling were sold to Bowring's.

By now, the 'dark clouds that had gathered over Europe' had brought with them the tragedy of the First World War. Expedition crew members in the *Terra Nova* and members of other expeditions from the heroic age joined the armed services and many did not survive.

A company record shows *Terra Nova* bringing in seals during the 1914 and 1915 seasons, when under the command of Captain William J. Bartlett and again in 1916 under the command of Captain S.R. Winsor, followed by Captain Nicholas J. Kennedy in 1917.

*Lloyds Register* confirms that the captain of *Terra Nova* between 1914 and 1916 was Captain W.J. Bartlett. In the 1916–17 register her captain is recorded as S.R. Windsor [Winsor], and from 1917 to 1919 she is recorded as being commanded by Captain N.J. Kennedy, who commenced his command in 1916.

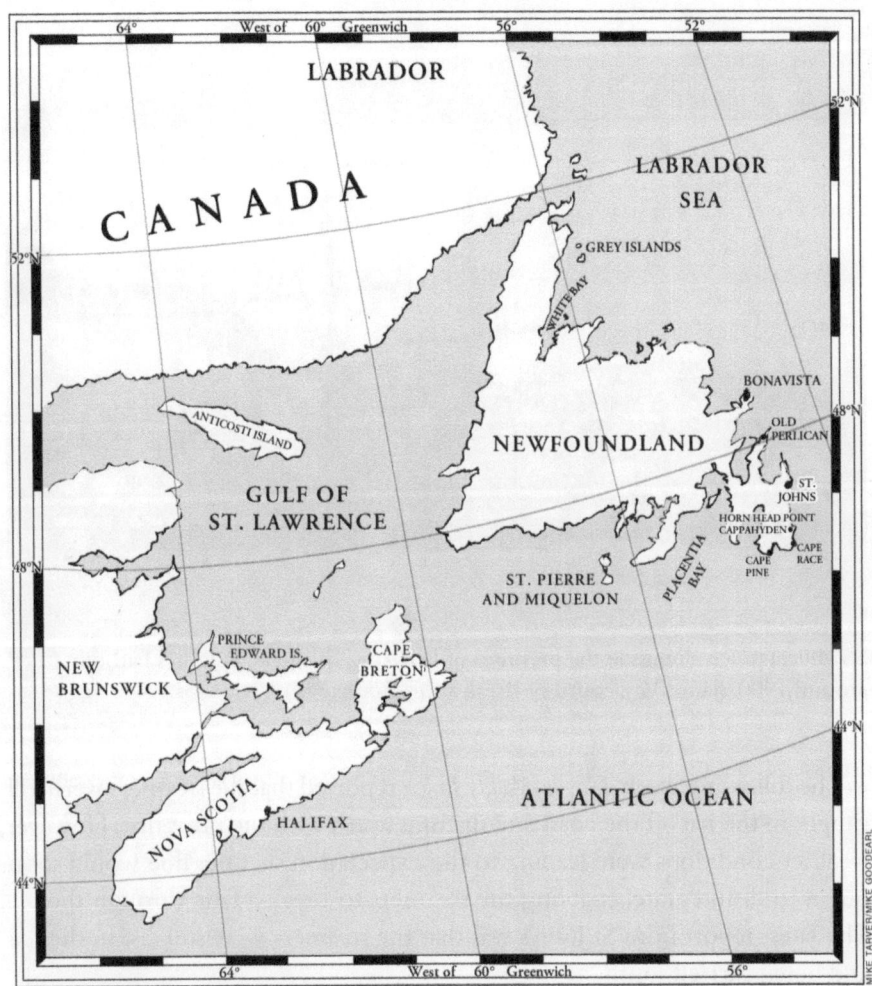

(Chart compiled by author)

*Lloyds Register* shows that from October to December 1915, she was undergoing 'conservation' (presumably a refit) and was therefore out of commission. But on 5 April 1914 she was at St John's, Newfoundland, having sailed from the Gulf of St Lawrence, and was still in that port on 31 January 1915.

On 8 March 1915, she sailed for the seal fishery but by 10 March it was being reported that the sealing steamers, *Terra Nova*, *Viking*, *Diana* and *Erik* were beset among ice floes 20 miles from St John's with nearly 700 men aboard. By 12 March 1915, it was reported that they had escaped serious disaster but the situation had looked so bleak that twenty men from *Erik*, fearing the wreck of their ship, had walked ashore across the pack ice.

SS *Aurora* in ice alongside the premises of Bowring Brothers, St John's harbour, around 1900. (James Vey, courtesy Breakwater Books Ltd)

The following week, *Lloyds Weekly Index* reported that the steamers were still caught in the ice off the coast and drifting south with the great floe. However, weather conditions were leading to the expectation that the floe would soon begin to disintegrate, enabling the steamers to break a lane through the ice. The latest report from St John's was that the steamers were still fast in the ice and being carried south.

The two men who had served as captains of *Terra Nova* during the 1910–13 Antarctic Expedition resumed their careers in the Royal Navy and had entered the war. Her navigating officer and relief captain, Lieutenant Harry L.L. Pennell, was promoted to commander and was navigating officer in the battle cruiser, HMS *Queen Mary*. On 31 May 1916, in a preliminary engagement which opened the Battle of Jutland, accurate firing from German warships, *Syditz* and *Derfflinger*, hit *Queen Mary*. A shell detonated her forward magazine and *Queen Mary* was blown in two. There were only nine survivors and 1,266 men were lost. Harry Pennell, much loved and respected member of Scott's last expedition, went down with his ship.

Captain E.R.G.R. Evans survived the war. On 20 April 1917, he was in command of the destroyer HMS *Broke*. In company with HMS *Swift*, she pursued a group of German destroyers which had bombarded Dover. In the engagement that followed off Calais, HMS *Broke* rammed and sank the German destroyer *G.42*.

Evans had a successful naval and public career. He retired from the navy in 1938 as Admiral of the Nore and after the Second World War entered the House of Lords as Baron Lord Mountevans of Chelsea. He died in 1957. It might be said that he achieved all the fame, professional and social position in life that Robert Falcon Scott might have wished for himself.

An archive letter of C.T. Bowring & Co. dated 16 July 1962 refers to the fitting of a new boiler and smoke stack in *Terra Nova*. Later photographs confirm the acquisition of a shorter replacement funnel fitted in *Terra Nova*, and it is thought that this was the funnel from the warship, HMS *Lobelia*.

HMS *Lobelia* was a fleet sweeping sloop for the Royal Navy, built in Scotland at the Clydeside yard of William Simons & Co. Ltd of Renfrew and launched on 7 March 1916. She was 1,250 tons and 268ft long overall and was one of ten Arabis Class warships built during the First World War. She was handed over to the Newfoundland Government in March 1920 and decommissioned in 1924. Photographs show the Arabis class ships fitted with two funnels, each shorter and stockier than the original smoke stack shown in earlier photographs of *Terra Nova*.

For the next thirty years, *Terra Nova* continued her duties in the North Atlantic sealing industry and as a coastal trader. The company memorandum also states that through the war years, 1914 to 1918, she was engaged by her company in transporting cargo from Canadian maritime ports to St John's and she made at least one crossing of the Atlantic to Cardiff with a load of pit props in about 1916, dodging the menace of U-boats. Research confirms this, and the *Cardiff & South Wales Journal of Commerce* records '*Terra Nova*' s.399, arrived at Queen Alexandra Dock, Cardiff, on 24 July 1916. From 26–31 July she was at the Roath Dock and sailed from the Queen Alexandra Dock in Cardiff on 1 August 1916. The name of her captain for this passage is not recorded and we can only refer to the list of captains for 1916, which shows that she was under the command of Captain Samuel R. Winsor, although Captain Nicholas J. Kennedy took over command in the same year.

## Rendering Assistance at a Maritime Disaster

During the early hours of Sunday, 24 February 1918, there occurred a company tragedy. The SS *Florizel*, a passenger steamer (3,081 tons) of the Bowring company's fleet, the Red Cross Line, was outward bound from St John's for Halifax and New York in a blinding storm of rain and sleet. Because of navigational errors, she came ashore 50 miles south of St John's on the coast near Cape Race at Horn Head Point, Cappahayden, and about 250 yards from the beach with seventy-eight passengers and sixty crew aboard. Her stern half was awash, with the sea battering her port side. It being a Sunday, there were difficulties in raising crews to man ships tied up in port at St John's to answer the SOS signals.

Eventually, *Terra Nova*, under the command of Captain Nicholas Kennedy, and four other vessels from St John's were sent down the coast to the scene. Heavy sea conditions made rescue attempts from the shore impossible and because of the conditions it was some hours before small boats could safely be sent across from the rescue fleet to the wrecked ship to take off survivors. Forty-four people were rescued and ninety-four were swept away from her decks.

Among those who lost their lives were Mr John Munn, a director of Bowring Brothers, his small daughter, Betty, and her nurse. It was a major disaster for the company. Ten members of *Terra Nova*'s crew and men of the other rescue ships, *Home*, *Hawk*, *Prospero* and *Gordon C*, were awarded Royal Humane Society Medals' for Bravery at Sea.

Between 1918 to 1923, *Terra Nova* is recorded as sealing under the command of Captain Abram Kean, overlapping with command under Captain Nicholas J. Kennedy. But on 23 April 1919 a speculative news item was published in *The Courier*, the Dundee newspaper serving central Scotland, announcing: 'Famous Dundee Vessel *Terra Nova* Chartered by a Noted Explorer'.

The article revealed that there was to be another expedition to the Antarctic regions and that it was already in the advanced stages of organisation. It was to be led by Mr John L. Cope FRGS. It declared that Mr Cope, a surgeon and biologist, was already well known as a member of the Ross Sea party of the Imperial Trans-Antarctic Expedition, 1914–17. He had survived at McMurdo Sound when the *Aurora* had been blown from her moorings.

The report went on to state that the expedition would depart the UK in June 1920 and would return after six years, having carried out an ambitious scientific programme. A London headquarters for the expedition had already been opened.

Discharging a cargo of seals at St John's. (Courtesy Tundra Books, Toronto)

Captain Abram Kean on the bridge. (Courtesy Tundra Books, Toronto)

It would seem, however, that after the First World War, as might be expected, there was little enthusiasm for such a venture and the seemingly overambitious project which included chartering *Terra Nova* did not materialise, failing to attract sufficient finance. However, John Cope did take an expedition to Graham Land on the Antarctic Peninsula the following year, with a somewhat controversial outcome.

For the year 1924, we have a personal account of a near disastrous season 'at the ice' for *Terra Nova* with Abram Kean in command. Three men were drowned and the ship was almost lost by the action of 'raftering ice'. The story is told in his 1934 autobiography (republished in 2000), *Old and Young Ahead*.

On the morning of departure from St John's, ice crammed the land outside the harbour. As the ship butted her way through ice towards open water, 3 miles out, three crewmen fell through a deceitful pan of ice and were drowned. Following the tragedy, it was two days before the ship reached open water and then steamed north toward land at White Bay before the ship became jammed and unable to move until the wind changed. The ship eventually made contact with seals in the afternoon.

The death of the three seamen is the subject of a song published in *Haulin' Rope & Gaff: Songs and Poetry in the History of the Newfoundland Seal Fishery* by Shannon Ryan and Larry Small. It is a song entitled '*Terra Nova*', collected in September 1967 by Dr Herbert Halpert and Samuel Fiander Jr from the singer, Mr Norman Payne of Ramea, Newfoundland. The loss of the men described in the song, although no year is given, ties in with the 1924 account by Captain Abram Kean in his autobiography. The men are named in the song and it is reproduced alongside this chapter:

The SS '*Terra Nova*'

One Monday morning March the Tenth,
It opened fine and clear,
The slob ice in the Narrows,
From St John's to Cape Spear.

But nothing daunted Captain Kean,
Upon the bridge he stood,
In the good ship '*Terra Nova*',
To search for Harp and Hood.

Three blasts came from her whistle,
Which woke up all the town,
And men from all directions,
To wharf came running down.

Her roll was called, all hands aboard,
A hearty lot of men,

Bound to the northern ice floes,
To fill her up again.

Another blast and then she hauled,
Away from Harvey's pier,
Her course out through the Narrows,
The helmsman straight did steer.

She hugged her way to the narrow head,
And soon was put about,
But the slob ice was getting hard
It was without a doubt.

With staysails set to catch the breeze,
She forced ahead full steam,
In half an hour the Captain said, We'll have Torbay abeam.

But very shortly after,
A pan got in her way,
Which caused the death of three bay boys,
Way out in St John's Bay.

A dozen of the pluckiest jumped on the ice that day,
To help the *Terra Nova* carry on her way,
But very shortly after,
Sad news it flashed to town,
That three of our brave fellows
In sight of all went down.

There was David Wheeler of St John's,
With a wife and children four,
Who was left without their bread winners,
It would make your heart feel sore.

There's Hubert Hiscock of Trinity Bay,
All taken in his prime,
His birth place sent out heroes,
To every land and clime.

And lastly comes young Mercy Breen,
Down from the southern shore,
Where the hardest boys are born and bred,
They've proved it o'er and o'er.

May the God of Love who reigns above,
Console those friends we pray,
Of those brave boys who gave their lives,
Out in St John's Bay.

[Published courtesy of Breakwater Books Ltd, St John's, Newfoundland.]

Abram Kean continues his story:

My men returned at dark and reported panning 5,000. That night the wind varied to the east-north-east and blew a gale. Next morning we had breakfast at daylight and I was up to give my men their instructions for the day, remarking to my officers how well we had escaped 'raftering' all through the spring. Kean went on ... A word of explanation is necessary in connection with 'raftering'. Large sheets of ice, larger and thicker ice driven by the force of currents and wind will double up and smash smaller sheets. When this takes place, it is called 'raftering'. It is especially dangerous when the ice is in large sheets and comes in contact with land; when the ice is broken in small pans by the lift of a swell, there will be no 'raftering.'

    Suddenly, I felt a nudge which I knew was a rafter. I dressed for the occasion for I knew it was stormy on deck. When I got on deck I met the doctor and one of my officers with their grips, leaving for the side of the ship. I was just about to stop them and upbraid them for their cowardice, when on reaching the bridge, I saw that about sixty of my crew, with their bags on the ice, had already left the ship. The ice was then coming over the rail and lodged by the galley door. This continued until I imagined there were ninety tons of ice on her deck. I watched the masts to see that they were in line and when the rafter stopped I heard from the engine room that the ship was not taking a drop of water. I then called for silence, telling the crew that there was no fear of their lives, that the ship was not making a drop of water and I named the masters of watch and gave them their work to do. By noon we had temporary repairs made, all the ice off deck and after dinner the crew panned some 5,000 more seals.

The next day we were still held in the rafter, the ice broke about a mile from us and my men could not cross. I then ordered them to launch four dories up to the rent for the crew to cross to the other side and to keep a crew with the dories in case other rents broke.

That day we panned altogether 17,000 seals. After that we could not reach the seals and were still held in the rafter, being driven fifteen miles from our pans. At the end of the ninth day we got clear, reached our pans and picked up our 17,000 seals and I don't know that we lost a single pan. Altogether we got 18,851.

Between 1925 and 1941, *Terra Nova* is shown variously under the command of Captains Abram Kean, Wesbury Kean, John Parsons, Richard Badcock and Stanley Barbour.

In 1931 another disaster hit Bowring Brothers. The SS *Viking*, a 'woodenwall' sealer of the fleet in which *Terra Nova* served, was destroyed by a sequence of explosions on board caused by the mishandling of gunpowder.

The *Viking* had put to sea with a film crew to photograph further action scenes for a new film to follow successful movies already made by the American producer, Varik Frissell, in association with a local photographer, Samuel R. Oakley. These films were called *The Lure of the Labrador* (1926) and *The Great Arctic Sea Hunt* (1927).

Oakley encouraged Frissell to make a new film, *The Viking*, and the script, a dramatisation of Newfoundland's traditional seal industry, was already written. Many scenes were shot in and around St John's and in the ice fields, where many sealers took part as actors. The principal actors and actresses were augmented by local people, among them Captain Bob Bartlett, another Newfoundland mariner who as a fisherman and sealer became famous for his Arctic exploration exploits in association with Robert Peary, Vilhjalmur Steffanson and their attempts to reach the North Pole.

When Frissell returned to the USA, he found Paramount Pictures were not satisfied with the footage and they required more action shots at the ice, including some of an iceberg capsizing. Frissell returned to the ice in 1931 aboard *Viking* under the command of Captain Abram Kean Jnr, with Captain Will Kennedy as the navigating captain. A total of 153 men were aboard and it was to be an authentic sealing voyage.

Extra gunpowder was carried for creating action scenes. While making her way to the ice fields near the Horse Islands, the *Viking* blew up and sank. Twenty-seven men including Frissell were killed.

It was long thought that the film was lost but when building work took place in 1950 at the old fishing company premises of Job Brothers in St John's, a copy of the film was found. In 1971, with the benefit of modern technology, the old nitrate film was copied and eventually it was put on video. It is a unique record of the seal fishery in the days of the wooden-wall steamers and the making of the film is considered to be one of the first sound picture films to be made while on location.

The post First World War years of the twenties and thirties were hard economic times. These hardships and the tragic losses suffered in the war left many households in a sad state. Conservation was not in fashion then, so *Terra Nova*'s famous role in polar exploration would have been forgotten. *Terra Nova* just laboured on in her life as a sealer, her famous past slipping away into history.

A list prepared from company records showing whaling and sealing, from *Terra Nova*'s launch in 1884 to her loss in 1943, records a total of 788,000 seals taken during those fifty-four years. Whaling was only carried out in the first eleven years of the ship's life from 1885 to 1896.

In a rough calculation made from the figures quoted, it appears that sixty-two seals make 1 ton. In 1898, the price per cwt quoted was $3.25 which amounts to $65.00 per ton. It can be said that one seal was worth less than a dollar in those days for its oil and skin. In a leaflet handed out to sealers getting their 'crops' in 1936, the price of seals hadn't increased much, reflecting the falling market. Young harps and young hoods were listed at $3.75 per cwt and old harps and old hoods were $2.50 per cwt.

The annual hunt for seals had been a way of life for Newfoundlanders for 150 years or more. It was a contributing factor to employment and the economy; not least, a way of life for those who participated in the annual season of the seal hunt, which began in March of each year and lasted only for two months. For the rest of the year, most ships spent their time tied up in St John's Harbour.

Up to the 1930s there was still a fleet of sealing vessels operating, principally consisting of the old 'wooden-walls' like *Terra Nova*, but these were joined later by some steel vessels. By 1941, only four sealers were still operating from St John's. Three of these, *Terra Nova*, *Eagle* and *Ranger*, were Bowring-owned, although *Ranger* had to be abandoned in 1942 after sustaining serious damage in a storm.

## Portrait of a Legendary Sealing Master: Captain Abram Kean OBE

A large island such as Newfoundland with a maritime history developed by its own people must produce many heroes. For more than 400 years, Newfoundlanders fished the Grand Banks and created a cod industry, and people of other nations came and settled, attracted by the employment and a way of life. They exported their natural resource to many countries and for the past 150 years the sealing industry provided employment, which became an important contribution to the island's economy.

Captain Abram Kean (1855–1945). (Courtesy Flanker Press Ltd, St John's)

The sea and fishing, ships and boatbuilding will always produce leaders of men, and Newfoundland had many famous captains of ships who led their men to the fishing grounds and to the dangerous icy conditions of the North Atlantic. They relied on these men, for many families they provided a few months of income when otherwise there would be nothing.

Out of this story emerges one iconic figure – Captain Abram Kean, who did not retire until he was 81 years of age and who died at the age of 90. *Terra Nova* was built as a sealer and no man had command of her more often in this role than Abram Kean.

He was born on 8 July 1855 at Flowers Island, Bonavista Bay, the youngest of nine children of a cod-fishing and sealing family whose parents could neither read nor write. His long life was certainly controversial. He knew both success and failure. There was tragedy at an early age and also later in life. All this moulded a man through life in a style which describes the very rugged life of the Newfoundland people and the island itself.

At the age of 10 he mishandled a gun, shot himself in the hand and killed his brother's 3-year-old son. He was the only one of the family to receive any education, but only for three years, and he was able to read. When his mother died he was just 13 years of age. As his sisters left home his father took in a young housekeeper, and when he was 17 Abram married her. She was seven years older than he. They had six sons and two daughters.

Eventually, with the death of his brother (the father of the 3-year-old boy) and the coming of a disease which crippled his third-eldest brother, followed by the death of that brother's wife, Abram took it upon himself to take his family under his wing. This was a burden which a close family in those days would feel it their duty to accept, there being no other help to hand.

In addition to his own family, he had an additional eleven more family members to care for. It shows Abram as the big-hearted and strong man that he was. He was a Methodist and deeply religious. His autobiography shows that he was a man of deep conviction with strongly held principles and opinions. As a man who had been fortunate enough to receive some education, he secured the services of two teachers and with the help of the Methodist Church established a school on Flowers Island.

Abram began work as a cod fisherman at the age of 13 and with his brothers worked the family schooner. He learned the skills of fishing and listened to stories of the sea told to him by his father's brother. This inspired him to make it a life for himself. Abram's father, however, by choice preferred the land-based activities of fish preparation. He moved his family from the island to the mainland, Abram gained marine qualifications and worked his way up to bridge master, sailing on a steamer.

In 1885, he entered politics and was elected to the Legislature. After three years, the family moved to St John's, where Abram qualified for his master's ticket. At the age of 34 he had his first command of a sealing vessel and made an impressive start with the SS *Wolf* to begin a career at the ice, and later he was in command of the coastal steamer service. His career lasted until his retirement in 1936 at the age of 81.

The annual sealing season was of two months' duration and a successful season brought with it a bounty, but there was time for other business during the year. Abram re-entered politics from 1897 to 1900 and became First Minister at the Department of Marine and Fisheries. He spent thirty-three years in the employment of Bowring Brothers Ltd, and again entered politics in 1927. As a captain and minister, his politics brought him into contact with those representing the interests of the fishermen, and there were many bitter conflicts.

In his business interests he does not seem to have been very successful. He lost money on ships and cargoes of fish and there are stories of poor business arrangements. Of his six sons, two became doctors and three qualified as ships' captains. But tragedy always lurked. The eldest son was drowned as a passenger, washed off the deck of the wrecked SS *Florizel*. His third son was drowned while ice skating on a pool near home when only 13 years of age.

During Abram's long career he commanded the steamers *Wolf, Hope, Aurora, Terra Nova, Florizel, Stephano, Nascopie, Thetis* and *Beothic*. He broke many records for the number of seal pelts brought into port, culminating in his final career total of 1,052,737, registered in 1934 near the time of his retirement.

Because of the financial rewards of his successful voyages, he was the person the men most wanted to sail with, but here again tragedy and bitterness lurked and affected his reputation. There was the *Greenland* tragedy in 1898, when forty-eight sealers found themselves adrift on the ice, separated from their ship which became frozen in and unable to reach them. An overnight vicious storm from the north took its toll, and the stranded sealers from the *Greenland* perished. It was not his ship, but it was suggested that the men were delayed on the ice by having to make up their pans, due to them being dishonestly taken by the crew of Abram Kean's ship.

Worse was to follow in 1914, with what was known as the 'Newfoundland Tragedy'. The *Newfoundland* was under the command of Westbury Kean, Abram's son. The fleet which went to the ice that year included the *Stephano* under Abram Kean.

On 9 March 1914 they left St John's, and south of the Grey Islands, north of Newfoundland, they began their work. However, *Newfoundland* became iced in, so the men were put overboard to work and told to make their way to *Stephano* where they would be directed as to where to find their quarry. After refreshment aboard *Stephano*, they were put overboard by Abram Kean and directed to a place where they would find seals. Captain Westbury Kean expected his men to stay aboard *Stephano* for the night, but Abram Kean expected the men to return to their own ship.

Instead, the wandering men spent two nights on the ice in bad weather, not able to get back to either ship or any other ship in the fleet and with each captain believing that the men were safe aboard the other's vessel. The main cause of the problem was the lack of communication. There was no wireless aboard SS *Newfoundland*, although the other ships were fitted with this new device. The owners, Harvey & Co. had removed it as being an unnecessary expense.

Abram Kean was later accused of putting the men back on the ice in bad weather and causing confusion when giving directional instructions. Seventy-seven men died in this incident and, to add to the tragedy, in that same month was the loss of the *Southern Cross* under Captain George Clark, with all 177 hands. The ship was last seen heading for St John's but was never heard of again.

Surprisingly, in his autobiography Abram Kean had campaigned for a restriction on the number of crew carried to the ice, but the politicians had not supported him. The priority was clearly the financial opportunities and rewards of sealing, which were greatly sought after but at a high cost to human life.

Abram Kean admitted he had a hasty temper, but was proud that he had cured himself of profanity at an early age. He firmly believed that in commanding over 100,000 people in his time, he had not found it necessary to use profanity. Indeed, men who were so inclined had asked for forgiveness in his presence and he believed that 'a man is to be pitied who cannot find enough words in the English language without descending to such vile substitutes'.

In 1934 Abram Kean had reached his target of bringing in more than 1 million seals. He had become known as 'Admiral of the Fleet'. A Flipper Supper* was held in his honour on 1 May 1934, attended by many dignitaries of St John's. He was presented with a model of the SS *Terra Nova* in a glass case, complete with a silk flag. When the King's Birthday Honours were announced on 3 June 1934, Abram Kean was awarded the OBE. He completed his autobiography in 1934, noting that Bowring Brothers had given the SS *Beothic* a thorough overhaul and, subject to continued good health, he hoped to command her for some years to come.

Captain Abram Kean retired from the sea in 1936 and died in 1945.

## A First-Hand Experience of 'The Greatest Hunt in the World'

Returning to the years of 1920, we are fortunate for the foresight of George Allan England, a US journalist, who specialised in unusual adventure stories. He wrote articles for *The Saturday Evening Post*, a popular American magazine of the time.

In 1922, George Allan England secured a passage in *Terra Nova* to see the seal hunt for himself and to tell the story that he described as the 'the greatest hunt in the world'. His account of the five weeks spent aboard *Terra Nova* in the sealing grounds north of Newfoundland was originally published in 1924 under the title *Vikings of the Ice*.

Without the five-week record of his experiences it would be difficult to portray the way of life aboard *Terra Nova* during her many years of this rough-and-tumble existence. England's story was republished in 1969 as *The Greatest Hunt in the World* and can be described as a documentary rather than just a story.

---

\* Seal flippers are considered a delicacy and are served as a traditional dish, particularly on social occasions.

'This veteran of the ice is dark, dingy, coal dusty and dirtier than anything I have ever seen.' (*Vikings of the Ice: The Greatest Hunt in the World* by George Allan England.) (Courtesy Tundra Books, Toronto)

When England first saw *Terra Nova* in 1922, he wrote:

> This veteran of the ice is dark, dingy, coal-dusty and dirtier than anything I have ever seen; with snowy decks, rusty old hand pumps, a stuffy and filthy cabin, extremely cold; tiny hard bunks, a dwarf stove, a table covered with smeared oilcloth; everything inexpressibly dreary and repellent, but that is comparative elegance to what she is to look like after four weeks of sealing. One could have kicked a hole in her boilers, and condemned; she should have been broken up for junk years ago. One must remember that not only did she go sealing that year but for another 20 years.

This is an account by someone who experienced life aboard *Terra Nova* in the principal role for which she was built. From England's vivid and descriptive narrative, I have put together sections of his story to give some idea of life aboard the ship and the character of the men working in her. In places, I have quoted his words exactly and in others I have summarised his work. Once again, I make no apologies for quoting his very words, for the story is his to tell and it is recommended to be read in full.

I am grateful to him for his foresight in recording the experiences for history so that we might enjoy his story, nearly 100 years later. It is so poignant, as it took place aboard SS *Terra Nova*, the 'flagship' of the Newfoundland sealing fleet, under the command of Captain Abram Kean. My thanks to Tundra Books of Montreal and Toronto, Canada, who in 1969 republished the story with an informative 'Introduction' by Ebbitt Cutler who acknowledged the story to Mrs Blanche Porter England Churchill, widow of George Allan England.

*Part 1: Joining the Ship*
No account of the *Terra Nova*'s sealing career can be bettered than that provided for history by George Allan England. An American author who specialised in explorer journalism and who wrote frequently in one of America's, at that time, leading magazines, *The Saturday Evening Post*. His book, *Vikings of the Ice*, was first published in 1924. His experience aboard the SS *Terra Nova* while on a five-week fact-finding trip to the Arctic sealing grounds presents a graphic account of life aboard the 'flagship' of the Newfoundland sealing fleet. At that time, only eight old ships, known as the 'wooden-walls' remained, where once there were hundreds.

England was born in Nebraska in 1877 and graduated from Harvard University with BA and MA degrees. His fascination with sealing brought him to Newfoundland to see for himself 'the greatest hunt in the world'. His story, *Vikings of the Ice*, remained out of print for more than forty years. It was republished in 1969 by Tundra Books of Montreal as *The Greatest Hunt in the World*, with an introduction by Ebbitt Cutler. In it is described 'all the blood, filth, waste and cruelty of the seal hunt and all the lust, poverty, courage and endurance of the hunters'. Ebbitt Cutler's introduction to England's story includes further information on the final hours of *Terra Nova* in 1943, as told by her last captain.

England had arranged a berth aboard the *Terra Nova* with her owners the Bowring Brothers, and was fortunate in that he was able to sail on one of the most romantic ships of the twentieth century under one of the greatest sealing captains of all time. *Terra Nova*'s part in polar exploration had already given her a place in history and Captain Abram Kean, who was in command for England's trip, had been more than fifty years 'to the ice' and had come in more than any other captain as 'high liner' of the fleet, holding records for total weight and number of seal skins brought in.

Sealing captain and mates in the old wardroom. Compare this photo with the wardroom photo on page 127. (Courtesy Tundra Books, Toronto)

In the annals of sealing, George Allan England's story remains unique. No one has ever written so revealingly, so richly, so indelibly of day-by-day life aboard a 'wooden-wall'. The story of *Terra Nova* would be incomplete without it and chosen extracts have been included in my *Terra Nova* story as a tribute to England's contribution to recording the life and work of those sealing men and in gratitude to the author's adventurous originality in journalism.

George Allan England's story began with a prefatory note dated 28 September 1923, in which he wrote:

> This book was written as the result of some six weeks' experience on two sealing steamers, the *Terra Nova* and the *Eagle*, out of St John's, Newfoundland. Its purpose is to fill a gap which has persisted astonishingly long. For many years the Newfoundland seal hunt has been the greatest hunt in the world and that so little has been written about it is a mystery. The world as a whole knows little of it. Even many Newfoundlanders of the 'better' class remain comparatively ignorant of this gorgeous epic of violence, hardship and bloodshed. In so far as personal observation can avail, I have tried to record and portray all the essential features of Newfoundland sealing.

He went on to make his acknowledgements and began with a short telegram from Mr Edward J. Penny, of Bowring Brothers Ltd, Newfoundland, who arranged the trip:

St John's, Nf
Feb 9, 1922.

Sealing trip arranged on Terra Nova with famous sealing captain. No passenger accommodation. You will be quartered with three junior officers. Be prepared to rough it.

Edward J. Penney.

England travelled from Halifax, Nova Scotia, to St John's in the *Rosalind*. He described the three-day lively trip, with gales, snowstorms and ice, finding St John's 'shivering' along her curve of icebound harbour, behind her mighty cliffs.

He went on to describe Bowring Brothers' establishment:

Bowring Brothers is a famous firm of seal hunters, and the *Terra Nova*, on which I was to ship, belonged to them.

I found Bowring Brothers' place of business comfortably old-fashioned, quite in the British tradition, with open fires, ancient desks, a one-handed English 'phone, maps, samples of seal oil, pictures of steamers, lots of clarks and the general atmosphere of 1848. Mr. Eric Bowring extended a cordial hand and reaffirmed his permission for me to 'go to the ice'.

He introduced me to Cap'n Abram Kean, scheduled to command the *Terra Nova*.

The Cap'n looked a splendid type of seaman and famous ice master, ruddy, hearty, hale, with shrewd blue eyes, a grizzle of snowy beard, a bluff manner, the vigour of a man of fifty for all his seventy years and a full half-century of seal killing to his credit.

'The Admiral of the Fleet' they call him in Newfoundland. And well he deserves the title, for he has come in 'high liner' more often than any other captain, and knows the icefields as other men know their palms. Many decades he has commanded ships plying to the far north and 'down the Labrador'. A skipper in the Royal Naval Reserve, a former member of the House of Assembly, a writer and lecturer, he understands more about seals and sealing than any other man alive. I thought myself fortunate in being assigned to his ship and so indeed the event was to prove.

In the slashing wind and bitter cold, England was taken to *Terra Nova* and his optimism faded on observing the scene before his eyes. There he saw the time-bitten old sealing ships of the 1922 hunt in the harbour in grinding white heavy ice. Beyond him snow-swept hills soared to a pitiless grey sky of storm. The wharves swarmed with types of men unknown to him, strange men, ominous and wild, with never a friendly glance or word for the outlander. Winches clattered and roared and steam drifted. After a look round he made his way back to the hotel, where dejection gripped him and he was tempted to give up, for he had 'cold feet' about it all.

The town was beginning to fill up with 'greasy jackets', the seal hunters, and he felt totally uneasy about the company he was to join. These men came from far and wide to join the spring hunt. Many had slept rough and walked in for miles or travelled in by the railroad, carrying with them their meagre bits of kit in ditty bags. They came in their hundreds, hoping for a berth. The captains gave out tickets and so did the officers. The tickets were eagerly sought as if some great price were being conferred, instead of a chance for week after week of unbelievable hardship, with just a few dollars' reward at the end:

> As *Terra Nova* and other ships of the fleet prepared for sea the wharves were crowded with expectant hunters and supplies going aboard. 'Cropping' is getting their personal equipment aboard. The sign 'Crop Here' directed where the men should get their sealing equipment, i.e. their tobacco, knives, boots or whatsoever they desired. The sealing companies allowed each man $9 worth of goods. If no seals were taken, the companies stood this as a dead loss; if seals were taken each man was charged $12 for his 'crop'. The chances of no seals being taken are about those of one's not seeing stars on a clear night.

As the ship loaded, England got to know the characters aboard, and he became known as the 'de quare fish of a 'Merikin', so quite clearly some explanation of the Newfoundland sealers' dialect is required.

As the *Terra Nova* prepared, he saw the *Viking* sail, with wives and sweethearts waving and cheering goodbye. The ship's whistle blared, and a tugboat broke the ice as she departed:

> 'Usa'll soon be gadderin' de pans now, b'ys, Soon 'aulin' em aburd o'dis un! Hello me son! First I t'ought it was y'rself, an' den I t'ought it was y' brudder, an' now I see 'tis needer one o'ye! Her'm 'aulin' off, now! Dere she go!'

Cap'n Bartlett bawled orders and the stout old Viking put to sea, swarming with weather beaten, patched, excited, greasy men. Tomorrow would be the day of *Terra Nova*'s departure.

The Newfoundland dialect of the time has been described as one of the most marvellous composites on earth. It is all of that. The Colony had been populated by English, Irish, French and Scotch and you can trace all these influences in the dialects.

Only a highly expert investigator could classify and tabulate them all correctly, with their various intermixtures.

## Part 2: The Seal Hunt

Before the ship departed for the ice, England explained the general idea of what the seal hunt was and where it took place and what the Atlantic herds were and how they lived and migrated.

He wrote:

The Atlantic seals are very different from their Pacific cousins. The latter have long curving flippers and commercially valuable fur. They maintain breeding colonies on St George and St Paul and on some of the other islands of the Pribilof Group (in the Bering Sea) and seem more distinctly land mammals than their eastern congeners. The seals in our aquariums and parks and those in vaudeville acts that bounce rubber balls from their noses and perform diabolical music on drums are usually Pacific ones. No self respecting Atlantic seal could be induced to enter an orchestra.

The great value of the sculps (skin and fat) or pelts, is due to the thick layer of pure white fat and to the extremely high grade of leather manufactured from the skin.

England went on to describe the sealing industry:

The Esquimaux first discovered the worth of this skin. They make skin boots, tobacco pouches, bags of various kinds, and many other useful articles out of this excellent leather. The Newfoundland seal hunters bring the skins to St John's every spring, hundreds of thousands of them; and there the skins are peeled from the fat. The skins are sent to England to be worked up into leather.

The fat, much pleasanter to read about than actually to smell, is ground, steam cooked, refined, sunned in glass roofed tans till it becomes a pure, white tasteless and odourless oil. And that a miracle, no less, for if anything in the world doesn't remind you of the *Roses of Gulistan*, it is seal fat.

Many are the uses of the seal oil derived therefrom. *My Lady Dainty*'s costliest soaps and perfumes often contain seal oil, and by chance her purest Italian olive oil holds a good percentage that which came from the frozen North. The finest of illuminating and lubricating oil, too, is a seal product. Purses, bicycle saddles, cigar cases, harnesses, binding for books are all made from seal leather. The value of oil and skins exported was a considerable amount to the economy of that country and the colony's prosperity hangs largely on the annual hunt. Next to codfish and its products, seal oil and skins give Newfoundland her greatest source of marine revenue.

The open season was generally 15th March to the lst May and coincides with the birth of the young, which bring a higher price because of the quality of the 'young fat'. Shooting the older seals is more expensive, because ammunition costs money.

The great danger aboard is as with all ships, fire, but especially so aboard a wooden ship creating seal oil which is about as inflammable as gasoline. It is almost impossible to extinguish, once it gets a fair start. It not only burns like mad, but it chokes and strangles its victims with thick black smoke. It is everywhere on a sealing steamer, such a ship well on fire becomes a hopeless case.

The sealing steamers are owned by a number of firms. The vast bulk of the hunt is carried on by the regular St John's fleet. Old wooden ships they all are, using both sail and steam and often carrying 160 or even 180 men apiece. 150 is a fair average. Built of green heart and oak, massively timbered with iron sheathed bows, these dauntless ships in charge of ice masters incredibly bold and skilful, slog out into the icefields and operate for five or six weeks between the Newfoundland and Greenland coasts.

The place of the hunt depends on how and where the vast annual migration of the *harps* and *hoods* happen to run. It is estimated that there are fifteen different species of seals, for the hunters two species are relevant, the *Phoca groenlandica* and the *Cystophora cristata*, i.e. respectively, the *harps* and *hoods*, the latter being the larger of the species.

In *The Bowring Story* by David Keir, writing of the Newfoundland seal hunt, the physical appearance of the two main species' harps and hoods are described:

> Those that are cream coloured when young develop a brown harp-sharped mark along the shoulders and sides when fully grown, by which time they are six feet long and weigh, on the average, a little over 300 pounds. The

larger hoods average 600–700 pounds and measure nine feet or more. Grey in colour, they are mottled with spots of brown; and they differ from the harps in other respects. The harps are gentle creatures, like their name. The hoods, on the other hand are aggressive and warlike.

They derive their name from the heavy folded sac of skin which the male seal inflates above his head when angered or alarmed; and indeed the sight of an old dog hood, blown with fury and thrusting his way forward in defence of his mate and her pup, is calculated to frighten the most intrepid hunter.

The season of the seal hunt has always been governed by the annual migration of the herds. Each autumn the harps and hoods leave their summer quarters within the Arctic circle for a 4,000-mile southward journey to the Grand Banks. The hoods from the east Greenland coast ride the ice floes which are swept towards the shores of Labrador by the Arctic current, to meet the harps drifting south from the east side of Baffin Bay. From there onwards the seals travel south in two long, roughly parallel columns, with the harps near the land. At the end of their journey they all reach the Grand Banks, where the rich supply of cod and other fish fattens them up for the return trip.

Early in February, the seals move north again, until they run once more into ice swept down to the east coast of Newfoundland by the Arctic current. The southern fringe of the floe, loose and broken, is known as 'slob' ice and it is through this that the seals make their way until they reach the heart of the Great Floe, where they congregate in hundreds of thousands for the whelping season. When the pups are less than a month old the seal hunt begins and shortly before that ... another migration takes place on shore ... the crews of the sealing fleets assemble.

The perils which confront these crews are often prodigious. In the sealing season the navigational hazards in this part of the North Atlantic are particularly severe. There is danger of violent north-east gales, with blinding snow and hail which cut down visibility to nil. Often the force of the gale whips the open sea into a swell which shatters vast fields of ice into broken pans and then, running counter to the tidal-race, piles the rafts of ice into crazy splintered masses jammed solidly against each other. There are also icebergs and huge low-lying ice chunks known as growlers, these are a constant menace to shipping. Fog and mist also abound and with a drop in temperature freeze over the decks and rigging and jam the running gear. Worst of all, the weather is never stable. There are odd days when tempest and blizzard are stilled and warm sunshine floods over the ice-fields from the clearest of blue skies, but any moment the glass may fall and a ship which has been running

Sealers at work on the ice. (Courtesy Tundra Books, Toronto)

easily through open lanes of water may find herself, only a few hours later, crashing into pack-ice or worse still, caught in a 'pinch', when a fresh gale unleashes its fury.

An enormous area is worked over. Day after day, week after week, the ships sometimes close together, often out of sight of one another, grind, crash, shudder through the ice, blast their way through it with bombs; drift with it when nipped, free themselves and struggle on against every possible obstacle and hardship that nature can fling against them. The ice will nip and sink a ship as easily as a nutcracker smashes a walnut, if pressures develop just right. Then there is to be considered the breaking of shafts and propellers and of labouring outworn engines, as has so often happend [sic], together with the possible explosion of boilers long since condemned [sic].

With a picture of what to expect, England prepared himself for the voyage aboard *Terra Nova*. On 8 March 1922, he boarded the ship with his only ditty bag and a battered suitcase. A coal fire roared below to dispel the clammy cold. Nobody had much to say but everybody stared, 'A writer eh? Why should anybody want to write or read anything about 'de hard rowt o'swilin?' He found himself accepted only under suspicion:

Cap'n Kean arrived, something of a fashion plate in a fine felt hat, well-cut overcoat and white collar, a fine old sea dog, proud, virile, dominant, and one of the real 'fore-now' men, which is to day the genuine old heart-of-oak breed of mariners.

England quoted from his diary:

> Gloomily, the old *Terra Nova* lay at her snowmuffled wharf as night comes on. On deck, sealers are carrying shovels of live coals to start a fire somewhere in preparation for a 'scoff' as a feed is called. Sparks are eddying from the aft-galley funnel. A stockish man, lumped down on a bench in the galley, is intoning a 'come-all-ye.' No sleeping accommodations, or any of whatsoever kind have been made for me. I have just dumped my 'fitout' and myself into a kind of little hellhole aft of the main cabin. This hellhole is partly occupied by the rudder trunk, partly by several rough black bunks. A tiny place it is, with a sign branded into a beam: 'Certified to accommodate one seaman'. Here I am awaiting developments. Everything is dim, dark, smoky, glum.

*Terra Nova*, her owners had decided, would not sail until the morning, so the evening was spent meeting the various characters aboard. The cabin swarmed with visitors till they gradually dwindled away and snores took possession. When morning broke, England felt refreshed and all hands tumbled out. Wearing a gigantic fur coat and hat, England was invited to the bridge by the 'Old Man':

> Vigorous tugs at the siren lanyard sent screeching echoes into the air over the sleeping town. Smoke poured from the galley and from the red crossbarred funnel. Gulls swung high, against a sky drifted with mottles of ominous black and cloudy purple. Then a reporter arrived with news that a steamer, the Grontoft had just sunk with all hands off Newfoundland. Encouraging, for a send off!

### Part 3: 'Goin' Swilin', is Ye, Sir'

'Four hands to the wheel!' the Cap'n shouted, for the *Terra Nova* was no luxurious, steam geared vessel. She steered by human muscle applied to twin wheels, in regular old-time style. On deck the jostle of men thickened. The gangplank was thrown off, Capt'n Kean wrenched the telegraph. The *Terra*

*Nova*'s engine, only 120 hp thrashed astern. It was 6.30 am on the morning of the 9th March as we headed towards the harbour entrance, then it would be north toward the Labrador Sea.

The *Terra Nova* swung nor'-nor'-east against a cold, strong wind ruffling frigid grey open waters to a sharp display of fangs and almost immediately slid into the wet wool of a fog that blotted away the towering headlands.

England was on his way north and received advice from various quarters on all aspects of the trip and the methods of taking seals. He told them that he wasn't going to hunt seals except with a notebook and camera, but it didn't sink in. He described his first dinner aboard *Terra Nova*, which came as a heaven-sent blessing. A brace of meats, white bread, lobscouse, turnips, potatoes and hardtack made a regal banquet:

The 1922 complement of the ship consisted of Captain, second hand, barrel men (or spy masters) scunners, engineers, firemen and oilers, bosun, carpenter, doctor, storekeeper, stewards, bakers and cooks, four bridge masters, Marconi man, master watches and second master watches (four of each), preacher, government inspector and common hands, a total of 160.

Roughly speaking, the proceeds of the trip are divided into three parts. One goes to the shipowners, one to the officers and one to the common hands. In general the one -third principle holds. The men's pay for the 1922, hunt according to figures compiled by Levi G. Chafe were as follows: *Sagona* $27.62; *Ranger* $18.13; *Thetis* $45.36; *Seal* $27.22; *Viking* $74.68; *Eagle* $49.22; *Neptune* $73.54; *Terra Nova* $74.90. The average pay was $42.40. Total net value of the catch was $197,837.91. If a hunter makes $50 or $60 he's doing well, that comes to $10 to $12 a week. [Levi G. Chafe, 1861–1942, kept statistical records of the seal fishery for sixty years as a hobby. He became employed as a customs officer on these matters and other duties relating to Newfoundland trade.]

Talk presently turned to perils and disasters, of not hearing the ship's whistle to return to the ship, of being caught drifting on the ice and of sudden fogs and wind changes. They talked of the Greenland tragedy in 1898 when forty-eight sealers were caught on the floes and died, and of the survivors who when rescued had gone mad and had to be dragged back to the ship.

Then there was the *Newfoundland* disaster in April 1914, when 119 men were caught on the ice by a north wind storm which raged for two days. Seventy-seven men died. The signal flag and siren from the ship went up, but the north wind strikes with fury and the warning was too late.

Around the same time in 1914, came another disaster, they recalled. The Norwegian-built *Southern Cross*, formerly used by Carsten Borchgrevink for the 1898–1900 Antarctic expedition, now commanded by Captain George Clarke went down with all 177 hands when returning from sealing. She was last seen from another ship on 31 March 1914 in open water off Cape Pine, not far from St Pierre Miquelon. Nothing more was ever seen of her, nor was any wreckage found, save a lifebelt picked up, months later, off the coast of Ireland. But still the sealing fleet went to the ice, no less courageously.

The *Sagona* lay astern of us as we ground slowly northward … '*wid a little better stick*' (speed) on us under steam and sail, the *Eagle* was racing us, off our starboard bow. Farther off, the Ranger came struggling. Who would first '*strike de fat*'? I began to share the nervous tension of the ship. It gets you, inevitably. That afternoon the glass took a decided upward turn and the sky cleared. The ice began to 'go abroad' a bit. Cap'n Kean went aloft. He came down, well pleased though vexed at the other ships so closely tagging. *Ranger*, *Neptune* and *Sagona* were identified astern.

'*They ahl want to dog in the same road*', he grumbled, his face like raw beef with frost and gale. He hated to admit the Newfoundland icefields were free for all. But anyhow he added … '*We made the field ice, now; got down to the reg'lar sheet ice, now, an' no more slob. You'll see somethin' like seal runnin' afore lang!*' He was radiant. '*We got great goin' now. Got a fine click on'er. A good tune on this one; she 'shakin' her tail some, now!*'

Chief 'Mac' beamed with equal satisfaction. He worried over his engine as if it had been a child; and now he revelled as she began to shove through the ice, grinding it to 'pummy'. A steel ship would have crumpled and split in a moment; but the *Terra Nova*, staunch veteran of nearly forty winters never even cracked.

Our first bit of news, early next morning Sunday, was the sighting of two huge seals off the starboard bow. Out on deck in haste, I beheld the immense old dog and bitch a couple of hundred feet away; and lively was the babel they raised. But, on account of the day, their lives were safe.

The decks were deep seated with snow – a dreary scene. Seals, safe for the day, were bobbing in a wide bay of water just astern. Whitecoats were bleating, some alone some in groups. A few were lying on their backs, wiggling their flippers in the most entertaining and debonair fashion. Not today should they die. Extraordinary, how the Sunday law holds! On the Sabbath men will work at anything but never will they kill. Some, however, draw the line at any kind of personal or avoidable work. I met hunters on the *Terra Nova* who refused to sew on a Sunday. 'De man above', they told me, would have regarded sabbatical needle and thread with disfavour. Of course 'chickers' were completely taboo. I never found any sealers, though, who carried the blue-law idea far enough to stop smoking on a Sunday or to refuse a good 'glutch o'rum.'

The bitch lay quiet on a gleaming pan, but the old dog, as the ship crashed by, kept turning round and round, then toddling forward with a sinuous, slithering and ungainly motion. I felt an odd sensation, lost in a frozen, insensate world, thus suddenly to come on life, and on such very warm-blooded, vertebrate, highly evolved life. I wanted silence. But no one else saw anything there but just two fat seals; and what a pity the day was Sunday! Men shouted gleefully: '*Dem de outscouts o' de patch, by's! Dis-un goin' rate direct fer de patch, jonnick (I swear) 'power o'swiles a'eed, I'm t'inkin. Good sign! Us got more luck 'an a cut cat!*' That afternoon the numbers of seals kept steadily increasing. Excitement ran ever higher. Alas, for it having been the Sabbath!

## Part 4: 'Not on the Sabbath'

Sunday night the ship lay idly rolling in slow heaving waters. For now was Church-time. Services were held below in the 'tweendecks. An unforgettable experience it was, to crouch down in the hold, to watch these Vikings at worship. To-morrow they will be glory butchers. This night they bow, with a sincerity that thrills you to the Power they feel very near. Every man pays devout heed. Some wear belts with the buckles of religious organizations. I can see on such buckles mottoes of 'Christian Lads' and 'Methodist Brigade'.

The vigour of their song is tremendous. Familiar old hymns, all. The preacher, Levi Butts, in spectacles and a brown sweater, leads them in song and prayer. He reads the Lord's Prayer, which all repeat. In their prayers and testimonies there are references to families at home and for far wanderers in the ice. In all their religion, as in all their lives, death ever obtrudes. These men think often of death. To them it is an ever-present contingency, not something far off and problematical, for every Newfoundland family has met sea losses. Death is no rare visitor as with us more sheltered folk. Not that they fear it . A braver, more heroic breed never lived. But – death is always there, just crouching.

That night we heard the whitecoats bawl and knew the kill was near. Early next morning it continued. *Thetis* and *Diana* were in the area too.

## Part 5: 'Old and Young Ahead'

On the following day, the interminable afternoon began to fag on, with naught to do save cough in the smoke-rank cabin and yawn, drunk tea, watch endless checker games; when all at once — A yell! From the masthead flung, it electrified our labouring ship: '*Whitecoats! Old and Young ahead.*'

Up tumbled all hands and out upon the coal-blackened decks. Spiked boots ground the planking. Forward, streams of hunters came milling from the to'gal'n house, the 'tweendecks, the dungeon. A rapid spate of cries, questions, cheers, troubled the frozen air. Grimed faces appeared at galleys, at the engine-room scuttle. Sealers lined the broad rails gesticulating out toward the illuminated plain of Arctic ice that blazed, dazzling white, under the March sun. Cap'n Kean had gained the bridge ... '*Overboard me sons*', he shouted '... *make a pierhead jump an' get into 'em! Over me darlin' b'ys!*'

But the men needed no urging, this was a free for all scramble. The kill was in full cry. Swiftly the men ran and leaped over the rough ice. They caught seals, struck with their heavy, cruelly pointed and hooked gaffs. Everywhere men were going into action. Everywhere the gaffs were rising, falling; tow ropes being cast off; sealers bending over their fat booty of both young seals and old. Everywhere the seals were being rolled over and sculped. Almost invariably the seals met death head on. They might flee at man's approach but once he was upon them, they would stand and show fight. The actual work of blood at first though later I grew used enough to it was rather shuddering to me. It was not a pretty sight to watch seals die.

If they could not be hauled aboard with gaffs, straps and ropes were used to winch them aboard. Straps were passed through a bunch of sculps and the 'wire' or rope from the winch dragged out from its pulley on a spar, by the whip-line, eager men hook the strap of seals to the wire. With a roar and rattle, a hissing of steam, the winch snakes up its quivering load. Shouting men tug at the whipline, holding back the sculps as best they can from catching on the side-sticks. Up, up the ship's side the sculps drag and then swing free, a heavy, dripping pendulum of hair, fat, skin and blood. They fall on the deck, while joyous hands grab, unhook, twitch out the now bright red whip-line and fling it all a sprawl once more far over the rail. The ship's first bit of wealth is *aburd o' dis-un*'. And so it all goes on until the kill draws to its close for lack of killable material.

'*Come ashore now, ahl hands!*' orders the Cap'n; the word 'ashore' in Newfoundland ship talk meaning 'aboard'. '*We're going on, now. Maybe gin' to get another rally 'fore night!*' Night is approaching. The west is beginning to flame with gold and scarlet. But still enough light may endure for a bit more slaughter. The men cheer and laugh as they swarm in. Up ropes and over side-sticks, red-painted now, they escalade with the agility of apes. They catch the rail with gaffs, haul themselves to the rail, leap over the reeking deck.

Easy 'starn! from the Old Man. The engine-room bell jangles. Out backs the *Terra Nova* from the bottom of the 'bay' where she had lain. The archaic engines begin to thud and thump again, like a tired heart. Away the ship surges, away from that red-blotched place of desolation where, save for some few frightened survivors still surging in sunset-lined waters, all seal life has vanished. The first 'whitecoat cut' has been made. Man has passed.

Away the ship grinds, crushes, shudders through the floe, but now with how exultant a spirit! Her men are different men. For the first honours of the spring are the *Terra Nova*'s. She is now, as till the end she remains, 'high-liner' of the fleet.

The *Terra Nova* bell in its original cradle acquired by Lieutenant Atkinson and later presented to the Scott Polar Research Institute, University of Cambridge. (Michael Tarver, 2005)

*Terra Nova* moved on to new killing fields. One night, England recalled lying in his bunk, with the ship 'garaged in ice', as was the practice, the engine throbbed all night to keep the propellor slowly turning to keep open water at the stern so as to be ice-free. At 5.30 a.m. he awoke, wondering what the hour was in 'bells'.* *Terra Nova* did not use 'bells', but told the time by the clock, as

---

\* The original bell from *Terra Nova* is at the Scott Polar Research Institute, Cambridge University, having been acquired by Surgeon Lieutenant Edward Atkinson, a member of Scott's last expedition. It is rung daily to announce the morning and afternoon tea break. It is not known whether the bell was replaced on the ship. The figurehead from the bow of *Terra Nova* was presented to the city of Cardiff by Mr Frederick Charles Bowring, before the ship finally departed destined for St John's in August 1913. The figurehead is now in the possession of the National Museum of Wales at Cardiff.

if ashore. He said that the only bell ever rung on board was the dinner bell, except for a metal bar hung in the engine room, now and again struck with a bolt, perhaps to try to keep up the nautical tradition.

Though he had gone on the trip joking to the sealers that he had only come with his notebook, pen and camera, England was anxious to earn the respect of the men and he took his turn at the work of hauling and dragging the catch across the decks, which he found exhausting work. He took his turn steering at the wheel, emptying ashes, peeling potatoes, counting seals, helping to load ice, a bit of doctoring and selling tobacco. He wanted to be seen earning his keep.

### Part 6: 'A Ship in Trouble'

News came one day by wireless that *Diana* was in trouble, jammed and unable to move with her propeller shaft broken and her bows 'bent in'. Later, more news came that the crippled Diana was beginning to have trouble with her crew and that mutiny was threatened. She still lay imprisoned in the ice, and most of her crew were beginning to demand relief from the other ships.

An urgent message came from the *Diana* demanding that a ship stand by and rescue her rebellious crew. Night brought news of the final scene in the crippled *Diana*'s career, a message that her crew had, in 'good earnest', abandoned her in a sinking condition and burned her with all her thousands of sculps still aboard.

The official account of this event, given by Levi G. Chafe in his invaluable, *Report of the Newfoundland Steam Sealing Fleet for 1922*, is brief:

> The 'Diana' killed and panned 7,000, and on the 16th March, the ship being jammed all day and while endeavouring to get clear, lost her tail shaft. She was abandoned on March 27th, about 100 miles S.E. of Cape Bonavista. The crew of 125 men was taken off her by the SS 'Sagona': Captain Job Knee; and landed at Old Perlican.

On his eventual return to St John's, England interviewed one of the *Diana*'s officers and went on to describe more fully what had happened. He learned that many of the sealers aboard were First World War veterans and, with the ship crippled and helpless, their spirits quickly rose up. They grabbed the firearms aboard and threatened the captain and radio operator, demanding that an SOS be sent. The ship was set on fire with all her thousands of sculps aboard. The crew left her burning and sinking, watching as they departed on their relief ship, the *Sagona*.

The *Eagle* alongside *Terra Nova*. (Courtesy Tundra Books, Toronto)

In his story, England pays tribute to the contribution made by the Newfoundlanders to the Allied forces in the First World War. Ten thousand Newfoundlanders enlisted in the army and navy and some 6,000 went overseas. They fought at the Dardanelles, Gallipoli, Suvla Bay, the Somme, Monchy-le-Preux, Gueudecourt, Passchendaele, Cambrai, Polygon Wood and Ypres. They served in the Royal Navy and Merchant Navy on North Sea convoy duties and perilous mine-sweeping work. The Newfoundlanders, like others, had sustained heavy losses. Many of the sealers had served in these situations, and seeing danger aboard the *Diana*, were quick to react.

## Part 7: 'Pierhead Jump for Home'

On 12 April, after five weeks aboard *Terra Nova*, England was told by Captain Kean that he would be taken off by the SS *Eagle*, under the command of Captain Edward Bishop. At 5.30 p.m. that day, word was sent down to him that *Eagle* was coming alongside and he would now be homeward bound. England dropped everything, dressed in his greatcoat and furs and made himself ready to make a 'pierhead jump' for the *Eagle*. He made his farewells and gratitude and saw the *Eagle*, her rails lined with the crew waiting to see 'dat quare fish, de 'Merikin'. The crew of *Terra Nova* shouted and waved their caps, wanting to be sent photographs. It was farewell to *Terra Nova*.

Aboard *Eagle*, he was welcomed by Captain Bishop and at 9.30 p.m. had sardines with his night mug. Life aboard was pretty much like *Terra Nova*. *Eagle* had rain, wind and fog all the way to St John's, with gulls trailing behind with their plaintive cries, no doubt scenting the cargo. As the harbour came up, England saw the Cabot Tower and the Sugarloaf Heights. It wouldn't be long now before he stood on solid earth. In through the Narrows, the ship's whistle shrieked, people waved and a British Man-o'-War lay at anchor. As *Eagle* tied up, England jumped ashore, which he described as a 'never to be forgotten' contact. His voyage with the 'Vikings of the Ice' was over and done with.

George Allan England's full story *Vikings of the Ice*, republished by Tundra Books in 1969 as *The Greatest Hunt in the World* is well worth reading. He ends his story with eight verses he wrote aboard the *Terra Nova* during a blizzard, in which he pays tribute and says farewell to the 'Sealers of Newfoundland'.

## THE SEALERS OF NEWFOUNDLAND

Ho! We be the Sealers of Newfoundland!
We clear from a snowy shore,
Out into the gale with our steam and sail,
Where tempest and tumult roar.
We battle the floe as we northward go,
North, from a frozen strand!
Through lead, through bay, we fight our way,
We Sealers of Newfoundland.

Yes, we be the Sealers of Newfoundland!
We laugh at the blinding dark;
We mock the wind as we fling behind
The wilderness, hoar and stark.
We jest at death, at the icy breath
Of the Pole, by the north lights spanned.
In a wild death dance we dice with chance,
We Sealers of Newfoundland!

Sealers ho! Sealers of Newfoundland,
With engines begrimed and racked,
With groaning beams where the blue ice gleams,
We push through the growlers packed.
With rifle, with knife we press our strife;

What blubber shall understand,
The war we fight in the ghostly night?
Aye, Sealers of Newfoundland!

The ice glows red where our skin boots tread,
And crimson the grinding floes.
From mast we scun till our race be run,
Where the Labrador current goes.
From ship we spring to the pans that swing;
By stalwarts our deck is manned.
O'er the blood-red road the sculps are towed
By the Sealers of Newfoundland.

Oh, some may sail with a southern gale,
Some may fare east or west.
The North is ours, where the white storm lowers,
Wild North that we love the best!
O North, we ken that ye make us men;
Your glory our eyes have scanned.
Hard men, we be, of the Frozen sea,
We Sealers of Newfoundland!

Bitterly bold through the stinging cold
We vanquish the naked North,
We make our kill with an iron will,
Where the great white frost storks forth.
'Onward!' we cry, where the bare bergs lie,
Dauntless our course is planned.
With blood, with sweat, scant bread we get,
We Sealers of Newfoundland.

'Starb'rd!' and 'Steady!' and 'Port!' we steer;
Press on through the grinding pan!
We labour and muck for a fling at luck,
Each of us, God! a man!
We cheer at the bawl of the whitecoats all,
We labour with knife and hand,
With rope and gaff. At the North we laugh,
We Sealers of Newfoundland!

Where the old dog hood and old harps brood
Lie out on the raftered pack,
We tally our prey. Then away and away,
Men, ho! for the homeward track!
Till the day dawns near when a rousing cheer
Shall greet us, as red we stand
On the decks that come to our Island home,
We Sealers of Newfoundland!

[Reproduced courtesy of Tundra Books, Montreal]

# 13

# Chartered for War Duties

### Refit and Role as a Coaster Trader

In the years following the First World War, *Terra Nova* and the 'wooden-walls' were worked hard 'at the ice' but spent much time out of the sealing season tied up in St John's Harbour. In 1938, *Terra Nova* underwent a refit in dry dock at St John's. Sheathing was renewed on her bow. Most of the timber beneath was said to be as sound as on the day she was built. When fitting a new oak stem plate to replace one originally fitted by the builders in 1884, a portion of the bow on the starboard side had to be cut away to remove the old stem. The original timbers, planking and sheathing and the heavy nails showed practically no sign of stress or wear.

In building the ship, manila rope was used for caulking instead of the oakum often used at this time and this too was found to be as good as it was fifty-three years previously, when it was driven home in the shipbuilding yard at Dundee. To supply this stem, a baulk of English oak 30ft long by 26in square had to be cut.

Photographs of *Terra Nova* at this time show that her bridge was moved to be positioned forward of the funnel, rather than behind the funnel as in her earlier photographs.

*Terra Nova* undergoing a refit at St John's, around 1938. (Courtesy Varrick F. Cox)

*Terra Nova* in St John's Harbour. Note the changed position of the bridge and the replacement funnel, around 1938. (Courtesy Bowring Brothers Ltd)

## On Charter During Wartime

As already stated, a company list shows *Terra Nova* active with the annual seal catch during seasons in the 1920s and 1930s, under the command of Captain Westbury Kean (son of Abram Kean), Captain Jacob Kean, (nephew of Abram Kean), Captain John Parsons and, up to the early 1940s, Captain Stanley Barbour and Captain B. Richard Badcock.

During the Second World War in July 1942, she was chartered by Bowring Brothers Ltd, to Newfoundland Base Contractors (US), a US company supplying military bases in Greenland.

All efforts have been made to trace the logbooks of the *Terra Nova* for various periods in her sixty-year life. But only those which relate to the two periods, 1903–04 and 1910–13, together with Crew Lists and Agreements for those years, have at the time of writing been traced and these are in the archives of the Scott Polar Research Institute, University of Cambridge. In May 1941, the offices of the parent company, C.T. Bowring & Company, India Buildings, Water Street, Liverpool, were bombed by the Luftwaffe and consequently most records were lost. Those records may have included the ship's logbooks.

Crew Lists and Agreements for periods from the 1920s to the late 1930s are available in the maritime history archive at the Memorial University and the Provincial Archives of Newfoundland and Labrador, but searches there and at other official sources, including the Registry of Shipping and Seaman at Cardiff and the Public Record Office, Kew, London, for the ship's logbooks during these periods and other years have so far been ineffective.

## Ice Damage to the Stern: The Last Voyage and an SOS Call

The end for *Terra Nova* came in September 1943. There are a number of accounts on record in books and publications which describe the loss of the ship, ranging from 'loss by fire' to being 'sunk by enemy action', but an authentic account of her loss was put together in 1996, some fifty-three years later, together with a verbal account given by an actual witness to her sinking off south-west Greenland on 13 September 1943.

An account written by the captain and later, the radio operator, have been published over the years. The last voyage of *Terra Nova* is told here in the order in which it came to the notice of the author.

*Terra Nova* had left St John's for Greenland on 29 May 1943, with an all-Newfoundland crew. In difficult conditions while delivering to Greenland ports during June and July the vessel struck ice and damage was caused to her stern.

A report in *Lloyd's Weekly Casualty Reports, 20–26th August, 1943* is recorded as follows:

> London, Aug. 19 The following cable has been received from the owners of British sealer *Terra Nova*, dated St John's NF., Aug. 17: American charterers advise *Terra Nova* extensively damaged and impossible to effect repairs to make her seaworthy for return voyage. Information presently very meagre but understand vessel now at Julianehaab and presumably has been or is ashore.

The following further report was received:

> Lloyd's Weekly Casualty Reports, Aug. 27 Sept. 2, 1943: St John's, NF., Aug. 26 A message from the master of the steamer *Terra Nova* stated that damage has been caused by ice while keeping vessel clear of land in heavy storm. Survey has been held by local shipwright and master of American auxiliary vessel *Colonel F. Armstrong* with assistance of diver. Damage consists of greater part of stem missing from 5 ft. below water-line, part ice sheathing on starboard bow missing, bow plates and forward ends of planking torn and broke, approximately 15 ft. of counter keel forward missing. Pumps were worked continuously to control leak. Port anchor lost but replaced. Temporary repairs effected by filling openings and fastening plank ends with aid of diver and filling with cement inside. Leak in engine-room well aft now 3 in. hourly, not making water amidships. Surveyor's and master's opinion now safe for vessel to return to St John's, NF., in ballast. Have telegraphed approval, subject to surveyors granting certificate of seaworthiness Lloyd's Agent per Salvage Association.

Temporary repairs were therefore made at the Greenland port of Julianehaab with the ship still afloat. She departed for St John's on 12 September 1943.

However, the following was received:

> *Lloyd's Weekly Casualty Reports Sept. 1016 1943* report: London, Sept. 13 The following wireless message was received at 11.10 pm GMT on September 12: 'S.O.S. Whaler *Terra Nova* Lat. 60° 15' N Long. 46° 48' W'

An extract from the ship's log of the US Coastguard Cutter (USCGC) *Atak* is in the archives of the Scott Polar Research Institute, Cambridge and entries dated 12 and 13 September 1943 deal with the loss of *Terra Nova* on her last voyage.

The USCGC *Atak* (242 tons) was the former trawler *Winchester*, built by the Bath Iron Works in Maine and launched in 1937 before being commissioned by the US Coastguard in 1942 with call sign 'WYP 163'. She was 128ft in length overall, diesel powered with a cruising speed of 9.5 knots. She was based with the Greenland Patrol at Boston, Massachusetts.

On the evening of 12 September 1943, the *Atak* was anchored at Simiutak Island, south-west Greenland and at 1920 hours she received an SOS distress call from *Terra Nova*. By 1930 hours she was under way from the Simiutak anchorage and making her way to Lat. 60° 15' N, Long. 46° 48' W, the reported position of *Terra Nova*.

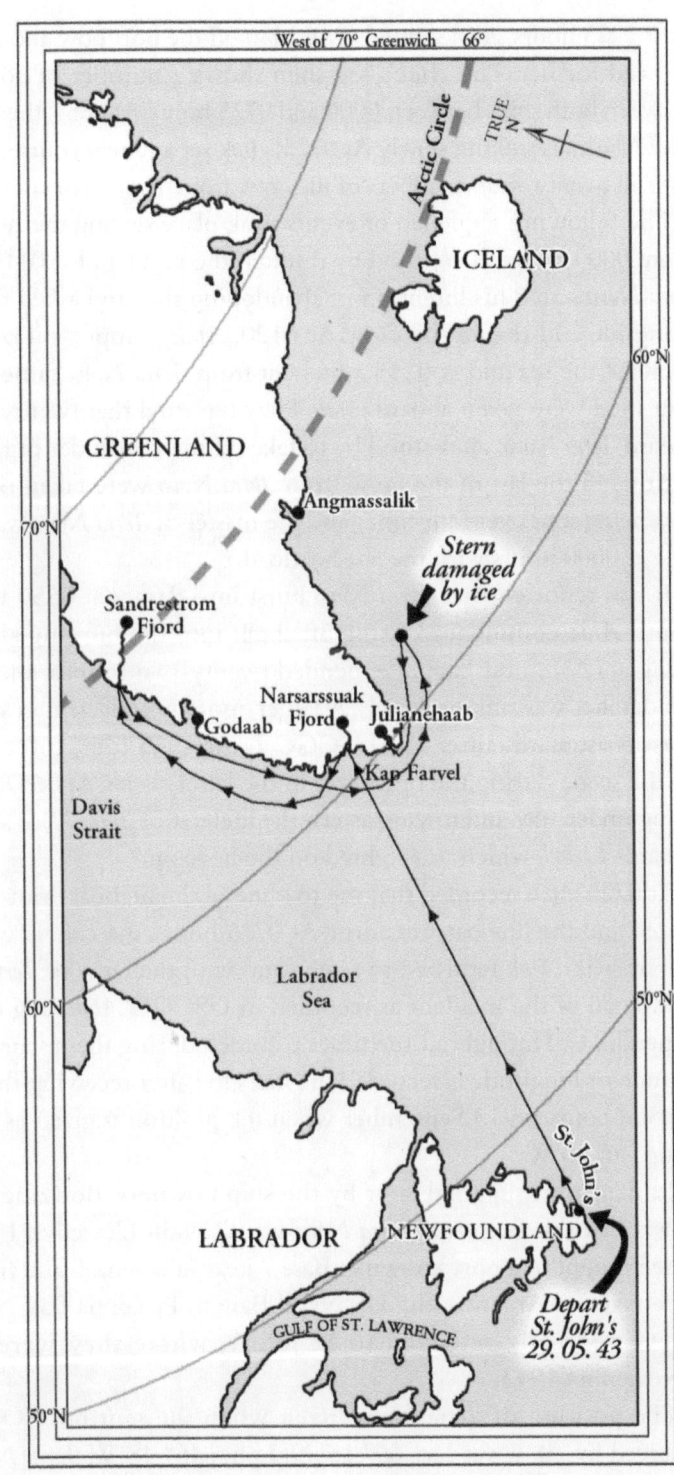

The last voyages of *Terra Nova*: (i) St John's to Narsarssuak Fjord, 29th May 1943; (ii) Narsarssuak Fjord to Sondrestrom Fjord, June 1943; (iii) Sondrestrom Fjord toward Angmassalik, July 1943, and after damage, passage to Julianehaab. (Chart compiled by author)

At 2225 hours, *Atak* sighted *Terra Nova* off the port bow and changed course to head for her. The *Atak*'s log then shows a number of communications between both ships between 0000 and 0725 hours. At 0655, the log records that *Terra Nova* was sinking slowly. At 0725, *Atak* set a return course for Narsarssuak with all twenty-four members of the crew from *Terra Nova* safely aboard.

The following sequence of events took place during the rescue operation. From 0000 hours, *Atak* stood by through the night and at 0115 the master of *Terra Nova* stated his intention of abandoning ship and asked that *Atak* come alongside and rescue the crew. At 0120, *Atak* pumped oil over the side to decrease the sea and at 0215 a lifeboat from *Terra Nova* came alongside and four men were taken aboard *Atak*. They reported that twenty men were left aboard *Terra Nova*. *Atak* stood by to take off the remainder of them.

At 0445 the last of the crew from *Terra Nova* were taken off, a boat from *Atak* being sent over four times and the master of *Terra Nova* was taken aboard *Atak* at 0600 hours with the last boatload.

It was reported that *Terra Nova* burst into flames at 0550 hours. At 0630 hours, *Atak* commenced firing 3ft shells into the burning ship and ceased firing at 0655, after having expended twenty-two rounds, and reported that the derelict was sinking slowly. A larger, more heavily armed vessel, a Cactus Class coastguard cutter, the USCGC *Laurel* (935 tons, 180ft overall), came on the scene during the operation in the latter stages. At 0657 *Atak* reported being under way and towing astern the lifeboat of *Terra Nova* and proceeding towards *Laurel*, which was sighted on the horizon.

At 0723 *Atak* recorded that the towline of the lifeboat from *Terra Nova* had parted and the lifeboat was adrift. At 0725 hours, the course was then set for Narsarssuak. *Atak* returned to Greenland with the crew of *Terra Nova* aboard. The detail of the incident as recorded in USCGC *Atak's* log can be seen in Appendix C. Throughout the times recorded during the rescue operation, no latitude or longitude is recorded in *Atak*'s log after receiving the SOS, except at 0800 hours on 13 September when the position is given as Lat. 60° 30' N, Long. 46° 30' W.

Information supplied later by the ship's owners, Bowring Brothers Ltd, shows that the master of *Terra Nova* was Captain Llewellyn Lush. The crew were brought to a port known as Base 7 near Julianehaab and from there they were taken to Argentia, the US Naval Base in Placentia Bay, Newfoundland. They eventually returned to St John's, where they were paid off on 2 November 1943.

The position of *Terra Nova*, given when she sent her SOS distress call received by *Atak* was Lat. 60° 15' N, Long. 46° 48' W. *Terra Nova* had gone

USCGC *Atak*, which responded to the SOS call from *Terra Nova*. (Courtesy Dept of Defense, US Navy)

about to return to the coast of Greenland after making this call. We must assume that the rescue activity took place in the proximity of the 0800 hours position, i.e. Lat. 60° 30' N, Long. 46° 30' W, on a reciprocal course from the position given by *Terra Nova* when the SOS call was made. Charts in use at the time and recorded in fathoms show the area with the depth at either 55 fathoms, or nearby 130 fathoms, Lowest Astronomical Tide.

It would have been necessary to ensure that, having been abandoned, *Terra Nova* sank and did not pose a navigational hazard. Tidal streams show that had she not been sent to the bottom, as a wooden ship she might have stayed afloat just on the surface and eventually drifted out to the Atlantic where, during wartime, North Atlantic conveys were operating.

Headlines of *The Evening Telegram*, St John's, Newfoundland, dated Thursday, 30 September 1943 reported: 'Veteran Sealer '*Terra Nova*' Ends Distinguished Career, Ship Lost in Arctic Waters, But Crew Are Safe Last but One of Wooden Walls She Saw Service in Antarctic Expeditions'. The newspaper report described much of the ship's history but, being wartime, no details of her sinking, position and the involvement of military rescue craft were reported.

## Memories and Recollections

The loss of *Terra Nova* on 13 September 1943 can be linked with further details which came to light in 1969. More information of the ship's loss was revealed with the republishing in 1969 of the story by George Allan England, *Vikings of the Ice: The Greatest Hunt in the World* by kind permission of the author's widow, Mrs Blanche Porter England Churchill.

New material was included by the publishers, Tundra Books Inc. of Montreal. It was republished with an introduction by Ebbitt Cutler in which he wrote about *Terra Nova* and quoted a letter he received a few months before publication of the new edition. It had been sent to him by Captain Llewellyn Lush of St John's, the last master of the SS *Terra Nova* in 1943. When he wrote his letter in 1969, Captain Lush would have been 79 years old, having been born in 1890.

Captain Lush described in his letter that last passage toward Greenland, the final voyage and loss of *Terra Nova*. His letter to Ebbitt Cutler told the story:

> I left St John's on May 28th 1943 [Saturday, 29 May 1943]* with a mixed cargo of cement and lumber. My crew were all from Newfoundland. I did not see land until I made Greenland and discharged cargo on the southwest coast. I was ordered to go on the east coast with cargo for the weather stations. I had some American personnel on board, working men, and had American convoys between the ports. All along the coast there was ice and numerous icebergs of the largest kind and dense fog most of the time. Back again to the south west, I encountered heavy ice and in order to keep her off the land and get the old ship through the heavy Arctic ice, I twisted her stern out below the water line. She started to leak somewhat, but the pumps kept against the leak and I got the old ship back to a place called Julianehaab.
>
> Divers went down and reported the old stern was gone right to the main keel. The divers patched the stern as much as they could There was not a dock to put the ship on and I was ordered to proceed to St John's.
>
> I left Greenland at 8 am, September 13th [Sunday, 12 September 1943]. At that time in the morning there was a moderate breeze, but as the day increased, it kept freshing to a stiff breeze and the Terra Nova commenced leaking badly. Sometime the same evening of the 13th [12th] she was leaking so badly the pumps could not keep against the water coming through the damaged part that the divers had patched up. I sent out the S.O.S about

---

\* NB: Time and date differences shown in brackets are corrected to correspond with incident log of USCGC *Atak*.

9 pm [7.20 pm] the same evening. The sea was getting rough at that time. Shortly after that the water got up to the fires, the dynamo went and pumps got choked. We were finished with regards of trying to keep any water out of the old ship.

There were some naval American ships around in that vicinity. About midnight I saw a ship blinking her dot and dashes. All we had was a flashlight to answer back. When the rescue ship got closer it was too rough, the sea too high, to be taken off. So I told the captain of the ship I would try and hang on until daylight. In the meantime the operator would keep in touch with him all night by flashflight.

At daylight all the crew got off by dory; yours truly and the first mate were the last to leave. Her decks were awash. She went to the bottom shortly afterwards. I saw the old ship sink. The rescue ship stood by until she went down and my crew and I were taken back to Greenland again to await convoy I would not like to be master of the old Terra Nova again at her age with a full load and heavy sea running.

Captain Lush added proudly of that final trip, 'I sat in the same chair at the table that the late Capt. Scott sat in his last voyage to the Antarctic'.

Efforts have been made through the media to trace descendants of Captain Llewellyn Lush and any surviving members of the crew, but without success. However, a further contribution to the story did surface in an article published in 1987 by *The Evening Telegram* of St John's. This article had been thoughtfully kept by Captain Tom Goodyear, a colleague of Captain Lush who died in 1981. Both had served as marine pilots for St John's Port Authority. Its content adds much to the story of *Terra Nova*'s final voyage and her last hours. Captain Tom Goodyear handed a copy of the newspaper article to the author on his visit to St John's in May 2005.

The 'Offbeat History' column of 6 July 1987, written by Michael Harrington, had told the story of *Terra Nova*'s history and this had been read by Mr W. Desmond Goff, a native of Carbonear, Newfoundland and now living in London, Ontario. Des Goff had been the radio operator of *Terra Nova* at the age of 21. He sent his own written account of the ship's last voyage to Greenland to Michael Harrington, and this was retold in 'Offbeat History' in the *Evening Telegram* of 21 September 1987.

When Des 'Sparks' Goff joined the ship, she was on charter to Newfoundland Base Contractors (USA). *Terra Nova* lay at the wharf of Bowring's Water Street premises in St John's. She was ready for her mission to take men and building materials to areas of Greenland for the building of meteorological and radio stations.

Captain Thomas H. Goodyear went aboard SS *Terra Nova* in 1935. At the age of 14, intent on making an early start to his seagoing career, he pipped off school and slipped aboard the sealing ships moored in St John's harbour. He made his way across to *Terra Nova* which put out into the harbour, where the afternoon was spent swinging the compass. He made his way below decks and remembers a cosy main cabin with a brass lamp. He recalled that he was inadequately dressed for that bitterly cold day and on the bridge he warmed his hands on the funnel which, he recalls, stood forward of the bridge. This is an interesting recollection which confirms the period of the refit, when in 1938 the position of the bridge was moved to stand forward of the funnel. Tom Goodyear joined the Furness Withy Shipping Line in 1937, obtained his deck certificates, worked his way up to master and served as a relief captain with the company. In 1952 he took up an appointment as a marine pilot with the St John's Harbour Authority upon the retirement of Captain Llewellyn Lush (the last captain of *Terra Nova*), whom he succeeded. By coincidence, their fathers served together in the RNVR during the First World War. Tom Goodyear retired in 1975. (Michael Tarver, 2005)

They were to assist the Allied war effort in detecting enemy aircraft, U-boats and surface warships. The harbour was full of warships and merchant vessels.

On Saturday, 29 May 1943, with her cargo of lumber and cement, *Terra Nova* steamed out of the Narrows. They progressed at 8 knots into a bitter north-east wind which gave them half-sleet and half-snow, with the dancing lights of the Aurora Borealis illuminating the scene. On the fourth day, the wind and sea dropped and they were into the remnants of Arctic ice calved from the glaciers. Goff wrote:

> It was disconcerting to feel and hear the ice sliding and crashing along the old hull, the sounds of her passage changing from the melodious noise of slipping through unobstructed water to the pounding and crunching of pushing her way through sluggish ice impeded waves.

Des Goff had written his account in a poetic form and his words were quoted:

> The seaward approach to Greenland is stunning and beautiful; the sight of the majestic land rising from a sea of frigid water, with the dazzling rays of

the sun reflecting from the vast icecap, forming a huge glittering plateau; the azure blue of the sky, the slight wind and high overhead a few seabirds floated effortlessly into the Arctic air, the icecap extending downward to the mouth of the huge glacier disgorging its abundance into the beautiful and stately fjord causes all to stand in awe and silence and behold this Greenland, often referred to as the great white stillness ...

*Terra Nova* made her way toward the village of Narsarssuak, on the north side of the fjord of the same name on the leeward side of the great icecap, and now the seas were calmer. The remaining chunks of ice growlers failed to impede her progress. The village was passed to the port side and was typical of most Greenland villages. There were fifteen to twenty brightly painted wooden houses along the shore and soon the ship was docked at the big US base, code named 'Bluie West One'.

An airfield located there was busy with military activity and *Terra Nova* remained in port for a week. It was mid-June 1943 before they made their way out to sea again, turned north and passed the Greenland capital, Godhaab and headed for Sondrestrom Fjord. They were now accompanied by two US Coastguard escorts which, from time to time, veered away over the horizon using their ASDIC and Sonar devices to search for enemy U-boats.

The next day *Terra Nova* closed with the fjord and made her way up to the huge base named *Bluie West Eight* which is 125 miles up the fjord from the open sea. The harbour was packed with warships and merchantmen, here *Terra Nova* docked and was loaded with materials for her next passage to the more forbidding east coast of Greenland.

On a crisp July morning, *Terra Nova* made her way down the fjord and turned south toward Cape Farewell (or Kap Farvel) on the southern tip of Greenland, and from there up the east coast to Angmassalik, almost on the Arctic Circle. There is no mention of any escort on this passage as they made their way around Cape Farewell and turned north.

There were savage intermittent gusts from the icecap on that long July day, as they pushed their way through an area of ice, which hindered progress. But it was in this area of ice that harm was done to her stern and all aboard heard the impact that caused the damage. Now she was taking on water and both pumps were brought into action.

She didn't reach Angmassalik but turned south and made it 100 miles back round Cape Farewell to Julianehaab. There, through August *Terra Nova* remained, but there were no dry dock facilities in that part of Greenland and a Danish diver had to go below the waterline daily with caulking compounds and other materials to try and get her ready for sea again. The diver had inadequate and obsolete tools and had to say that he could do no more. All that could be done was to wrap a huge tarpaulin sheet around the hull from No. 2 hold aft to the engine room area and secure it with ropes to the deckrails.

Captain Lush was anxious to proceed but his chief officer, Hansen, and the chief engineer (name not recalled by Goff) pleaded with him not to sail without an escort. However, the calls went unheeded.

Des Goff remembers departing on the crisp and clear morning of Sunday, 12 September 1943 with a nervous crew. *Terra Nova*, without an escort, put to sea for the last time. The crew had enjoyed pleasant weeks in Julianehaab while repairs were attempted but now their thoughts were full of anxiety with a 500-mile passage before them to St John's. Des Goff, the 21-year-old radio operator, gave no departure time but estimated that after a full day's steaming as nightfall came, they had put Greenland 100 miles behind them and he described the seas as 'getting up'. No speed of the ship or wind direction is given in Goff's story but if she steamed at between 6 and 8 knots then his estimation of distance covered would be correct. Captain Lush wrote in his letter that they had departed at 8 a.m. that day.

As the day passed the seas got bigger. The ship rolled from side to side, climbing up and rushing down the long seas of the North Atlantic. What they all feared then happened – the canvas sheet was torn from her hull and a call came from the engine room that the sea was rushing in through the damaged hull.

The engine room pump was started but soon jammed with coal dust and oil. The deck pumps were manned but quite unable to cope against the sea. The captain gave the order to 'Sparks' to put out an SOS giving their position and condition. After a time a reply was received that 'help was on the way', but now the question was on their minds – could they stay afloat long enough?

However, there is an inevitable difference in the recollections of what happened, which is only to be expected, considering the circumstances at the time and the period of years over which their stories are retold. The log of the USCGC *Atak* records that the SOS was received from *Terra Nova* at 1920 hours, giving her position as Lat. 60° 15' N, Long. 46° 48' W, which is a position 40 miles from Julianehaab.

After making the distress call, *Terra Nova* would have turned and made what speed she could toward the help which she hoped was coming. So between 1920 hours and the first meeting with *Atak* she must still have been able to make headway to cover the distance toward the shore of Greenland and the position at which both ships met.

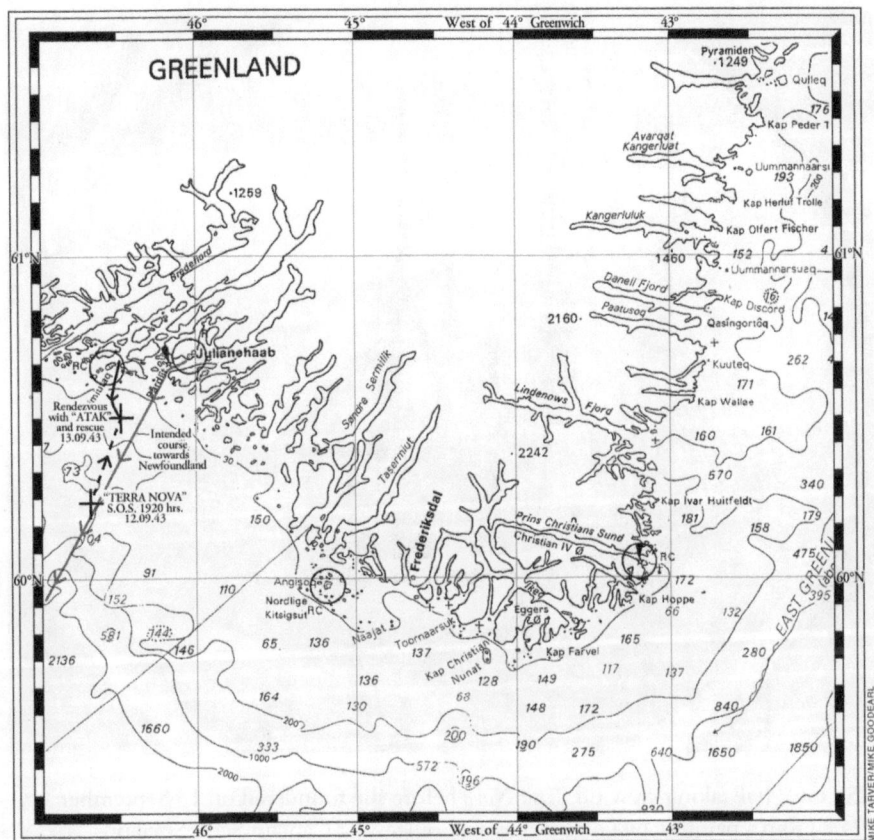

SOS rescue of crew and the loss of *Terra Nova*, Admiralty Chart 4405 (International Series). (Courtesy HM Hydrographic Office, Taunton)

If the SOS call was received by *Atak* at 1920 hours, this suggests that *Terra Nova* had been at sea for just over eleven hours (from 8 a.m., as stated in Captain Lush's letter to Ebitt Cutler in 1969, to 7.20 p.m., the time that the SOS signal was received).

The rescue ship flashed an Aldis lamp as it approached *Terra Nova*. Goff read its Morse signal, 'Help is on the way'. *Terra Nova* did not have an Aldis lamp aboard so Goff was sent to the bridge, being the only person able to interpret Morse messages, and all he had was an ordinary flashlight! Because of the conditions, he was unable to send signals and hold on to the bridge at the same time, so a seaman stood by and held him firmly to prevent him falling overboard.

Des Goff was in the lifeboat's last load to leave and had remained on *Terra Nova* to maintain communication with *Atak*, which stood off a few hundred

USCGC *Atak* taking crew off *Terra Nova* before she foundered on 13 September 1943. From a painting by Gus Swenson. (Courtesy of Captain Scott Society)

yards to leeward throughout the night. In the early morning, the sky brightened, the wind dropped and *Atak* moved to windward and was able to launch a lifeboat. Goff wrote praising the American sailors and described the difficult task they had in coming alongside the sinking *Terra Nova* and keeping the lifeboat away from her hull. On each rising wave, men from *Terra Nova* leapt from her rail into the rescue craft.

Goff maintains that *Terra Nova* was not set on fire, but that shells from *Atak* had acted as a *coup de grâce*. However, this account is at variance with *Atak*'s log, which records that *Terra Nova* burst into flames at 0550 hours. This is forty minutes before the firing of shells by *Atak* which is logged as commencing at 0630 hours. All crew and master were recorded as safely aboard Atak at 0600 hours, ten minutes after *Terra Nova* burst into flames, although the logging of this fact at 0550 hours is set down out of chronological order.

Goff described the final moments of *Terra Nova*:

As if in agony her bow reared up; she seemed to hang at about a 70-degree angle for one long moment. Then, as though unable to bear the burden any longer she slid swiftly below the waves she had lived on and known so well for almost 60 years. The rescue ship was so close that both crews could see the frothing whirlpool that marked her grave with small waves in a widening circle, that even reached and touched *Atak* as though giving a last farewell to all on board.

## Research and More Recollections

The following entry is recorded in *Lloyd's Weekly Casualty Reports* of 17–23 September 1943:

> St John's, NF., Sept. 14 Sealing steamer *Terra Nova*: A certificate of seaworthiness for the return voyage was granted by surveyors and vessel left on return voyage on Sept. 12. A message from the U.S. Navy now states that the vessel sprang a leak off Simiutak Island, South Greenland, and sank at [4.25 p.m.] GMT on Sept. 13. According to information received, pumps were unable to control the leak and boiler fires were put out by water. The vessel later sank after the crew were taken off by an American coastguard vessel. Lloyd's Agent per Salvage Association.'

The life story and aftermath of a famous ship will always be told and retold as each interested generation looks into history and discovers more facts. Accounts given by Captain Llewellyn Lush and Wireless Operator Desmond Goff came to light in 1969 and 1987 respectively, many years after the loss of *Terra Nova*. In 1996 an account of her loss was described, this time as viewed from USCGC *Atak* and given by her first officer (see later).

Ebbitt Cutler's introduction, which included the letter from Captain Lush giving his account of *Terra Nova*'s last voyage, also mentions the fate of two other ships that operated with *Terra Nova* on the seal hunts out of St John's. *Viking* suffered an explosion aboard in 1931 with the loss of twenty-eight men, plus another man from the ship who died soon after being rescued from an ice floe, and *Eagle*, the ship on which George Allan England returned to St John's after his passage on *Terra Nova* and the last of the 'wooden-walls' to survive, was taken outside St John's Harbour in 1950. She was no longer considered seaworthy by her owners, Bowring's, and was set on fire with the sea-cocks open.

Ebbitt Cutler added to his 1969 introduction:

> Ironically, a far different fate awaited *Discovery*, the ship Scott had used on his first expedition. After periods of idleness and an extensive refit in 1923, she did research work in Antarctic whaling grounds until 1931, when she was laid up in honourable retirement in the East India Dock in London. She now serves as a recruiting ship for the Royal Naval Reserve. Last summer, I was one of the 300,000 who visit her annually. As I walked through the beautifully refurbished cabins, with the mahogany and brass shining from Navy spit and polish, I thought of the *Terra Nova* disintegrating from old age and work in Arctic waters and I wondered which ending Scott would have more admired.

*Above*: Royal Research Ship *Discovery* and Exhibition Centre, Discovery Point, Dundee. (Dundee Heritage Trust)

*Left*: Sea Scouts aloft aboard *Discovery* around 1946 at King's Reach, Embankment, River Thames, together with HQS *Wellington*, HMS *President* and the majestic dome of St Paul's Cathedral. (The Scout Association)

The RRS *Discovery* did visit Barry Docks and then Cardiff* in November 1917 and Cardiff again in April and June 1918 during cargo voyages to French ports. She made her last First World War crossing of the Atlantic to Montreal and Hudson Bay from Cardiff in June 1918, under Captain G.H. Mead.

The *Discovery* was the expedition vessel for the two scientific Antarctic expeditions** known as BANZARE (British, Australian, New Zealand Antarctic Research Expeditions), led by Sir Douglas Mawson. The first BANZARE was from 1929–30 with the *Discovery* under the command of Captain John King Davis. For this expedition she took on 350 tons of coal and a quantity of compressed briquettes from the Crown Patent Fuel Co. at Roath Dock, Cardiff. Captain John King Davis had been with Shackleton on the *Nimrod* expedition. He had also commanded *Aurora* on the Australasian Antarctic Expedition, 1911–14, led by Sir Douglas Mawson. He was no stranger to Cardiff as he had taken on coal for *Aurora* in August 1911, on their way south for Mawson's Antarctic Expedition.

From 1936 to 1955, *Discovery* was managed by the Boy Scout Association as a training ship and accommodation for Sea Scouts visiting London, after which she was taken over by the Admiralty and then the Maritime Trust.

In 1986, *Discovery* returned to Dundee and is now docked at Discovery Quay, alongside a public exhibition centre at Discovery Point, managed by the Dundee Heritage Trust.

The loss of *Terra Nova* in 1943 is still not the end of the story because, fifty-three years later, the historical research group of the Captain Scott Society were researching the history of the ship. The society had formed the Terra Nova Trust in 1994 as a registered charity to launch a project to build a full-sized reproduction of the *Terra Nova* as an expedition and adventure ship. The plan was to build the ship in steel and she would be powered by diesel engines. She would have been a total replica in size and design. Presentation plans of the ship were prepared.

It was an ambitious project, estimated to cost around £12 million. The project was launched at Cardiff on the 12 June 1994 by Sir Vivian Fuchs FRS, President of the Captain Scott Society. Applications were made for National

---

\* It is from these visits that the Discovery Inn, opened in the early 1960s near Roath Park Lake, Cardiff, can attribute its name.

\*\* For the second BANZARE, 1930–31, *Discovery* was commanded by Captain Kenneth N. McKenzie.

## Chartered for War Duties

The foyer to the Captain Scott Room, Royal Hotel, Cardiff, 1996. (Courtesy of the Captain Scott Society)

Lottery and enterprise funding. The project was pursued for six years but, regrettably, wound up in the year 2000 as the funds were not forthcoming.

It was during research into the history of the original ship in support of the Terra Nova Trust project that information on the loss of the ship was provided by the US Coastguard Combat Veterans Association Greenland Patrol. Through them, the executive officer of USCGC *Atak*, Lieutenant Folke E. Swenson USCG (Retd), a resident of California, was traced by the historical research group. Folke Swenson vividly remembered the incident off Greenland in September 1943.

On 13 June 1996, at the annual dinner of the Captain Scott Society, held in the Royal Hotel, Cardiff, an eye-witness account of the loss of *Terra Nova* was given to those assembled by Folke E. Swenson, then aged 77. The society were re-enacting, as they do each year on 13 June, the last dinner given to Captain Scott, his officers and scientists before they had departed from Cardiff for the Antarctic eighty-six years previously. The event was taking place in the same banqueting room, at that time virtually unchanged since the day of the original dinner. The identical menu was being served, as it always is, and an orchestra played music from the original menu and programme of that day in 1910.

Lieutenant Swenson took part in the rescue of *Terra Nova*'s crew. From the deck of Atak he saw *Terra Nova* go down. In his speech to the assembled company, he gave a vivid account of the incident and said that all aboard *Atak* were aware that *Terra Nova* was a ship with a famous past.

Folke Swenson was accompanied at the dinner by his son Gus, a marine artist. He made a presentation to the Captain Scott Society of an oil painting by Gus, depicting those last moments in *Terra Nova*'s sixty-year life and the rescue of her crew. Gus's painting of the incident was the scene as recollected through the eyes of his father almost fifty-three years before.

In the year 2000, as a contribution to events to mark the centenary of the heroic age of exploration in the Antarctic, the Captain Scott Society launched an initiative supported by the UK Antarctic Heritage Trust to place a monument immediately on the quayside beside the dock entrance from where *Terra Nova* departed from Cardiff in 1910 and to where she returned in 1913. The monument, in the form of an iceberg embodying the faces of five explorers, was unveiled on 6 June 2003, by HRH the Princess Royal. Nearby stands the Norwegian Church, originally built in the nineteenth century at Cardiff Docks for visiting Scandinavian seamen. Both structures, as they stand together, are a poignant reminder of those events of yesteryear. Nearby is the Cardiff Opera House, home to the Welsh National Opera Company, and in the vicinity overlooking Cardiff Bay is a pub and restaurant appropriately named the SS *Terra Nova*.

When coming to the end of research for this whole story yet another surprise in the tale of SS *Terra Nova* came to light. It had long been thought that the figurehead of the ship in Cardiff and the bell at Cambridge were the only artefacts of the ship to survive. However, while examining a file in the archives of the Scott Polar Research Institute, the author came across a letter from Lieutenant Commander Robert Moss MA BSc, an instructor at the Royal Naval School of Navigation, HMS *Dryad*, dated 11 November 1944. This letter '... informed the Director of the Scott Polar Research Institute that the binnacle of the SS *Terra Nova* was at the Navigation School'. It being wartime,

The Antarctic Memorial and Norwegian Church, Cardiff Bay. (Michael Tarver, 2003)

the letter was acknowledged and placed in a file. News of the ship's loss in 1943 had probably prompted the letter.

Consequently, correspondence by the author revealed that indeed, the binnacle was at Portsmouth in the care of the Royal Navy Trophy Trust at HMS *Nelson*. This was the main standard compass binnacle fitted in 1910 to the roof of the ice house on *Terra Nova* and supplied by Messrs. Kelvin & James White 16–20 Cambridge Street, Glasgow and London.

This was the compass at which the navigating officer, Pennell, stood when he called out to the helmsman for a 'STEADY'. It had been presented to HMS *Dryad* in 1920 by Captain E.R.G.R. Evans.

Affixed to the binnacle is an inscribed plate, dated 1913, with the names of Commander Evans, Commander H.L.L. Pennell and Lieutenant H.E. de P. Rennick.

As one would expect, it has been maintained in an immaculate condition and the Royal Navy Trophy Trust kindly agreed on the recommendation of the Honourable Edward Broke Evans, son of Admiral Lord Mountevans of the Broke that the new home for the binnacle should now be the city of Cardiff, home port of the British Antarctic Expedition, 1910–13, where the binnacle will join the figurehead of *Terra Nova* for the public to view.

SS *Terra Nova*'s standard compass and binnacle. British Antarctic Expedition, 1910–13. Re-enactment of the 1910 photograph taken aboard *Terra Nova* at West India Dock, London. Left to right: Staff Officers Lieutenant Tim Davey, Lieutenant Tim Broughton, Lieutenant Mal Instone. (Courtesy of the commodore, Britannia Royal Naval College, Dartmouth. Photo: Michael Tarver, 2005)

# 14

# THE FINDING OF THE WRECK

The loss of any ship with such a famous past might seem to be the end of a story, but the finding of *Terra Nova*'s wreck off south-west Greenland now presents another chapter to add to the story of this famous ship. The locating of *Terra Nova*, especially in the centenary year of Scott's journey to the South Pole, is an important and exciting addition to the heroic age of polar exploration. Further opportunities as technology advances may open up to add to this enduring story.

A Cardiff man, Leighton Rolley, was working as a maritime research engineer with the Schmidt Ocean Institute based at Woods Hole, Massachusetts, USA. Rolley had always been inspired by the story of the *Terra Nova*. While aboard the Schmidt Ocean Institute Research Vessel *Falkor*, under the command of Bernd Buchner, a previous captain of National Geographic ships *Endeavour* and *Hanse Explorer*, they realised they both had a passion for the *Terra Nova* story. The search for the wreck provided the opportunity to test newly installed equipment aboard the RV *Falkor* while the vessel was in the North Atlantic on other duties.

But let us first go back to the Second World War, which saw increased shipping activity in the North Atlantic. The war saw the growth of the German submarine fleet specifically to interrupt the supply of urgent foodstuffs brought by merchant shipping to Britain and occupied countries in Europe. Then followed the supply of military arms and equipment to Russia to assist in the Second Front. The USA also had military air force bases in Greenland which were supplied by sea. The US Coastguard Service operated a fleet of armed vessels in the North Atlantic in addition to the roles performed by the Canadian Coastguard Service. These vessels, as well as being armed with guns on deck, carried depth charges and sonar equipment designed to combat the German submarine menace.

You will remember that after being acquired by the Bowring company in 1898, *Terra Nova*, operating between Newfoundland and Scotland, was drawn into the heroic age of polar exploration. She was purchased by the Admiralty to act as a relief ship to assist Scott's *Discovery* expedition, 1901–04, and then on her return to Britain, she was acquired by the American millionaire, William Ziegler, to relieve the Fiala–Ziegler Arctic Expedition, 1903–05, who had lost their ship crushed by ice. After a brief few years back in Newfoundland, *Terra Nova* was acquired for Scott's 1910 expedition which became known as Scott's *Terra Nova* Antarctic Expedition, 1910–13.

After fourteen years of polar activity in which she sailed from the Arctic to the Antarctic, *Terra Nova* returned to her owners, Bowring Brothers in St John's, Newfoundland, in 1913. From here she operated for the next thirty years, serving under many different captains and performing duties as a coastal trader along east American Atlantic ports. She also crossed the Atlantic to Britain on more than one occasion.

The year 1939 saw the outbreak of the Second World War which brought about increased shipping activity in the North Atlantic. It was into this theatre of war in July 1942 that *Terra Nova* was chartered by her owners to Newfoundland Base Contractors, a US company set up in St John's to supply military bases in Greenland.

The account of these times can be found in Chapter 13, which describes the last voyage of *Terra Nova*. She left St John's, Newfoundland, for Greenland, a 500-mile passage, on 29 May 1943. Her crew were Newfoundland men under the command of Captain Llewellyn Lush. She was carrying what is described as a mixed cargo of cement and lumber for maintenance of the US Air Force bases in Greenland.

Further north, her presence was monitored by armed US Coastguard vessels while she made her way along the south-west coast of Greenland and up the long fjords to make her deliveries. On her passage down the south-west coast and back up the north-east coast of Greenland toward the Arctic Circle, she struck heavy ice and was seriously damaged to her stern below the water-line.

# The Finding of the Wreck

They managed to return and enter the port of Julianehaab* where divers went down and carried out repairs as best they could. After a few days, decisions were made that the ship could return to sea.

The following sequence of events, which occurred on 12 September and during the early hours of 13 September 1943, have been put together from archive documents supplied to the Schmidt Ocean Institute by the US Coastguard Service and received aboard the RV *Falkor*. They were compared with information already made known and published in 2006 in my book *SS Terra Nova (1884–1943): Whaler, Sealer and Polar Exploration Ship* (ISBN 09552208-0-7).

At 0800 hours on 12 September 1943, *Terra Nova* left Greenland for the 500-mile return passage to St John's. Later that day, the wind and sea increased but the pumps were unable to cope with the ingress of water around the repair damage. The captain put out an SOS at 1930 hours that evening, when they were about 40 miles out from the Greenland coast. Four US vessels responded to the SOS call, *Atak*, *Laurel*, *Manitou* and *Amarok*. *Terra Nova*'s well-known past seems to have been of knowledge to many and by those who went to her aid.

*Atak* was the first ship to reach *Terra Nova* at midnight and circled the distressed vessel:

| | |
|---|---|
| 0120 hours | *Atak* received request from master of *Terra Nova*, requesting *Atak* to come alongside and take off the crew. |
| 0215 hours | Lifeboat with 4 crew from *Terra Nova* alongside *Atak*, reporting 20 men still aboard *Terra Nova*. |
| 0445 hours | Boat sent over 4 times to pick up rest of the crew. Master of *Terra Nova* aboard with last boat load. |
| 0550 hours | *Atak* reporting that *Terra Nova* had burst into flames. |
| 0630 hours | *Atak* reported commencing firing 3' 23 into burning derelict – 22 rounds expended. |
| 0700 hours | USS *Laurel* on scene. Lifeboat of *Terra Nova* being towed by *Atak*. |
| 0710 hours | *Laurel* reports Schooner [illegible] found afire and sinking. All her crew taken off by *Atak*. Approximate position [------redacted---------] |
| 0722 hours | *Laurel* reports all guns ready 3' 50 gun crew prepared to fire on schooner to remove navigation hazard. |
| 0723 hours | *Atak* reports tow line parted and boat adrift. |

---

* Now Qaqortoq.

| | |
|---|---|
| 0724 hours | *Laurel* reports firing on sinking schooner – 13 rounds expended. |
| 0725 hours | *Atak* sets course for Narsarssuak. |
| 0820 hours | *Laurel* reports lifting lifeboat aboard. |
| 0920 hours | Laurel reports firing 120 rounds of 20mm into sides of sinking schooner. |
| 1200 hours | *Laurel* reports *Manitou* and *Amarok* arrived at scene. |
| 1212 hours | *Amarok* instructed by *Laurel* to proceed to Nassuak, presence not now required. |
| 1310 hours | *Manitou* reporting wreck sinking …… fathoms. |
| 1325 hours | *Laurel* recording position. RDF (Radio Direction Finder) bearing of [not disclosed] from Simiutak Island. |
| 1433 hours | *Laurel* departs towards Simiutak Island, *Manitou* following astern. |

Those events took place thirty years after *Terra Nova*'s involvement in polar exploration but they were not going to be the end of the story. Still a fascinating subject to many and an enduring combination of polar exploration and maritime history, the loss of the ship brought excitement to the minds of those with a passion for such past events, particularly with modern developments and advances in underwater research technology. All this presented an opportunity to those aboard RV *Falkor* to follow up the story with the use of the up-to-date equipment to hand.

The Schmidt Ocean Institute is a non-profitmaking private foundation focussed on oceanography and founded by US citizens, Eric and Wendy Schmidt in 2009. The institute's goal is to advance ocean exploration, discovery and knowledge using technological advances, data-rich observation and analysis and open sharing of information. They are not wreck hunters.

USCGC *Laurel* also responded to the SOS call from *Terra Nova*. A 'Catcus' class tender, 935 tons, 180ft o.a. diesel powered, built by the Zenith Dredge Co., Duluth, Minnesota. (Courtesy James FIghting Ships of World War Two)

Initially, they operated a vessel named *Lone Ranger* until 2012, when she was replaced by the Research Vessel *Falkor*, a 272ft former German Fisheries protection vessel purchased from the German Government. RV *Falkor* underwent an extensive refit at Peters Schiffbau shipyard in Germany to convert to a globally capable oceanographic research vessel.

In 2012, with the refit completed, *Falkor* left Germany, via Newcastle in the UK and Nuuk in Greenland, for the USA. Opportunities were considered to test various items of the new equipment aboard on the way, one of which was surveying the southern coast of Greenland. Working in this location obviously brought the attention of marine technician, Leighton Rolley, to the location of the wreck of SS *Terra Nova*, which had gone down off south-west Greenland.

Research was already being carried out on sea-floor conditions. With all the topographical considerations and the secondary possibility of using a wreck as a calibration reference for the sonar equipment, the Schmidt Ocean Institute prioritised this location as the optimal spot for this round of tests.

As anticipated, numerous iceberg strikes and gouges were observed on the seabed, along with striking features not listed on the existing nautical charts. The contrast between hard and soft sediment signatures exceeded expectations. Gullies and gauges had collected soft sediment, while the surrounding flat seabed consisted largely of gravel and coarse material deposited by icebergs and glaciers with clearly contrasting backscatter.

On the first line of the calibration survey, Leighton noted a feature on the seabed which remained initially unidentified. Upon completion of the main calibration exercise, Leighton and Jonathan Beaudoin, from the University of New Hampshire, reviewed each of the many potential targets identified during the twelve hours of surveying and the target was noted as a strong candidate for further investigation. Multibeam data expert Jean Marie from Ifremer (Ocean Research France) analysed the feature in more detail, finding its length (57m) to match the reported length of *Terra Nova*.

Encouraged by the similarity in length, the acoustic survey team post-processed the collected multi-beam data to verify the observed feature. A shorter survey from several angles reaffirmed the possibility that the team had found a wreck.

During the earlier states of the transatlantic cruise, the Schmidt Ocean Institute technicians had an opportunity to develop a weighted camera package to film the plankton net trawls conducted by scientists aboard RV *Falkor* from Woods Hole Oceanographic Institution. This footage was designed for outreach work to show how these nets capture plankton which our scientists then study.

The camera package, named SHRIMP (Single-molecule High Resolution IMaging Package), is a solid metal frame with two attached cameras and three to four dive flashlights to document the planned plankton net tows. As the observed feature lay within the depth range of SHRIMP, the Schmidt Ocean Institute team decided to use the camera to take a closer look.

The package was dropped to a position just above the target to help identify the nature of the 57m-long feature observed in the EM710 output mapping data. Camera tows across the top of the target showed the remains of a wooden wreck laying on the seabed. The camera footage also identified the funnel of the vessel next to the wreck. The forecastle of the vessel appeared to be 'peeled' backwards to the port side and at an angle from the rest of the ship. The team compared the funnel image with historical photographs of the SS *Terra Nova*. All observations jointly identified this wreck as the sunken SS *Terra Nova*.

The finding of the lost SS *Terra Nova*, one of the most famous polar exploration vessels, was an exciting achievement – in addition, serving to successfully verify the performance and operational condition of the Schmidt Ocean Institute's RV *Falkor* multibeam echo sounders. It was made possible thanks to the collaboration and exhaustive efforts of all those onboard.

Schmidt Ocean Institute offers its profound thanks to the sonar experts and scientific representatives from the University of New Hampshire, Ifremer and the Woods Hole Oceanographic Institution for their excellent support throughout this exercise.

In July 2012, the Schmidt Ocean Institute announced that RV *Falkor* had found the exact location of the *Terra Nova*, off south-west Greenland, the ship which had carried Captain Scott and his expedition to Antarctica in 1910. Further research of the incident from US archives provided to RV *Falkor* successfully led to the discovering the wreck's precise location and as being further east than previously thought based on information previously provided. The GPS precise position of the wreck remains confidential to the Schmidt Ocean Institute.

(Taken from the report prepared by Leighton Rolley, aboard RV *Falkor* for the Schmidt Ocean Institute, 25 October 2012.)

# Epilogue

By the end of the Second World War, demand for seal products had fallen away and trade had declined since those early days. *Terra Nova* had begun her life as a whaler and sealer. She had seen the whale trade decline in Scotland and now the seal trade was in decline in Newfoundland.

It was the seal hunt which had laid the basis of the Bowring company's nineteenth-century success. *Terra Nova* had played her part in that, too. Now, few of those old ships were still engaged in the fishery and the great tradition of the old 'wooden-walls' was fading away. No longer did crowds turn out when the fleet left St John's to cheer them on their way, nor to lay bets on the size of the catch and the name of the first ship home.

The age of whaling was gone, the sealing fishery was in decline. Improved shipbuilding technology had overtaken the steam-powered wooden whalers and sealers which had shared their duties with polar exploration. The heroic age had passed and given way to the establishment of scientific bases and expeditions in the furtherance of science. The ships of those days were now part of history. *Aurora, Algerine, Endurance, Esquimaux, Diana, Erik, Kite, Morning, Neptune, Nimrod, Ranger, Scotia, Southern Cross, Thetis, Viking* and many others, and now *Terra Nova*. They had all, one way or other, met their fate. By 1950 just one ship remained, the *Eagle*.

SS *Eagle* was the third ship of that name to sail under the Bowring flag. Built in Norway in 1902, she was brought to St John's the following year and sailed on her first trip to the seal fishery in 1904 and made, in all, forty-five sealing expeditions. She was acquired by the Admiralty in 1944 and seconded for

Model of SS *Terra Nova* at the ice showing sealers at work. It is on display at the Nova Scotia Museum of the Atlantic, Halifax, and came from the offices of Bowring Brothers. The model is undated and the builder not named, but examination of its detail suggests that it depicts the vessel at work in the period between 1898 and 1909. (Photo: Michael Tarver, 2005, courtesy of the Museum of the Atlantic, Nova Scotia)

duty in the Antarctic under the command of Captain Robert C. Sheppard, in support of other vessels to consolidate British bases in the Antarctic Peninsula region. This she did successfully and returned to Newfoundland in 1945. The *Eagle* had been part of Operation Tabarin, the British Government's action to maintain sovereignty over regions in the southern hemisphere during the Second World War, known as the Falkland Island Dependencies.

Captain Robert Sheppard MBE was the harbour master at St John's and had been a square-rigger captain. He lived at Fort Amherst where he and his family ran a teashop for visitors. The adventurous role that the *Eagle* played in Operation Tabarin is told in *SS Eagle The Secret Mission 1944–1945* by Harold Squires, who was her radio operator on that voyage.

As a tribute to her long service and her place in history as the last of the 'wooden-walls', Bowring Brothers Ltd brought the career of the *Eagle* to a striking close as they considered her no longer seaworthy. It was a farewell with full ceremony and honour. On the 23 July 1950, she was towed slowly down St John's Harbour, through the Narrows and out to the Cordelia Deeps beyond. She was dressed overall in flags and bunting and the sirens

The Narrows. Entrance to St John's Harbour from Signal Hill, looking toward Fort Amherst lighthouse where once stood a fort dating back to the early nineteenth century. Entry leading lights in line, 276° (true), Canadian Hydrographic Service Chart 1941. (Michael Tarver, 2005)

and whistles of all ships in the harbour sounded a last salute of farewell as she passed.

A crowd of thousands, including men who had sailed with *Eagle* since she first arrived at St John's, gathered on Signal Hill to watch her last passage and a swarm of small boats followed in her wake as she made her way out to the open sea. Here, at Signal Hill, Marconi's first transatlantic radio message from Cornwall, UK, in 1901 had been received and on the opposite side of the Narrows, the difficult entrance to St John's Harbour, stood Fort Amherst.

At the Deeps, a skeleton crew set the old ship on fire, opened the sea-cocks and knocked the coverings off the scuttling holes cut in the hull. This last duty performed, they were taken off by the pilot boat and *Eagle* slowly settled in the water. With the Red Ensign flying aft, the Bowring House flag nailed to her foremast and with smoke billowing from her hull, she went down with all the dignity of any ancient Viking funeral. Her last moments marked the end of an era.

Sealing as an industry described in the lifetime of *Terra Nova* has diminished and today an annual quota of 300,000 is imposed on the catch for the Gulf of St Lawrence region. This story tells of a ship built for an industry of yesteryear, embodying a way of life created to harvest animals in the name of the fishing industry, which today has strict quotas imposed with all its consequences.

Sealing and fishing was the foundation of Newfoundland's economy in those days but now the industry is in conflict with conservation groups and their interests seem irreconcilable. Seals take vast quantities of fish, but they bite out only the nutritious parts leaving the rest to waste. It is estimated that each seal can take more than a barrel of fish a day while the fishermen have to operate within their stipulated fish and seal quotas and are left to ponder on competition from the seals.

With her sister ships, *Terra Nova* lies at the bottom of the sea in her North Atlantic grave beneath the ocean on which she was built to serve. She did so faithfully for sixty years in her many roles, with the honour of having traversed those seas from the Arctic regions, across the world to the continent of Antarctica.

It is worth wondering that, had *Terra Nova* not foundered in September 1943, would she have been part of Operation Tabarin and made one last return to the Antarctic? At over 60 years old, would she have been chosen over *Eagle*, which was 40 years old? Would sentiment have played a part in the choice of vessel? We shall never know.

Perhaps her loss in 1943 denied her just one more voyage to the Antarctic and a last moment of glory. But she had already gained for herself a place, not only in maritime history, but an important place in the scientific story of extending the boundaries of our knowledge towards of the polar extremities of the Earth.

# Appendix A

## Ships Built at Dundee by Alexander Stephen & Sons 1844–93

Ships built by Alexander Stephen & Sons, at Dundee

| Yard No. | Type of Ship | Name of Ship | Dimensions | Builders' Old Tonnage (Gross*) | Built for | Owners' Port | Completed |
|---|---|---|---|---|---|---|---|
| 1 | Brig | "Diana" | 86.2' × 22.3' × 14.3' | 191 | A. Ilives | Dundee | 1844 |
| 2 | Schooner | "Jules" | 70.3' × 19.9' × 11.7' | 123 | Baxter Bros. | " | " |
| 3 | Barque | "Brechin Castle" | 115.3' × 25.8' × 17.5' | 371 | " | " | 1845 |
| 4 | " | "Richard Cobden" | 116.6' × 25.11' × 17.8' | 361 | Wm. Small | " | " |
| 5 | Snow | "Catheeren" | 86.6' × 22.2' × 14.6' | 192 | Mr. Fenwick | " | 1846 |
| 6 | Barque | "Queen" | 115.3' × 25.8' × 17.5' | 370 | D. Martin & Co. | " | " |
| 7 | Snow | "William" | 87.2' × 22.5' × 14.7' | 197 | Scott & Murdo | " | " |
| 8 | Brig | "Neva" | 88.8' × 23.5' × 14.5' | 218 | Andrew Low | " | 1847 |
| 9 | Snow | "Jean Andson" | 85.8' × 21.0' × 12.6' | 172 | D. Martin & Co. | Arbroath | " |
| 10 | Barque | "Asin" | 105.5' × 23.6' × 15.6' | 274 | Andson & Duncan | Dundee | 1848 |
| 11 | " | "Europe" (Troop Ship "Dudbrook") | 139.0' × 25.5' × 10.8' | 548 | Alex. Stephen & Sons | London | " |
| 12 | " | | 143.6' × 28.6' × 20.4' | 551 | Mr. Mann | | |
| 13 | Brig | "Duna" | 93.2' × 23.1' × 14.6' | 228 | A. Low, Junr. | Dundee | 1849 |
| 14 | Ship | "Amazon" | 144.0' × 31.6' × 21.0' | 791 | J. & F. Somes | London | 1850 |
| 15 | " | "Cossipore" | 148.7' × 28.8' × 21.0' | 838 | W. S. Lindsay | " | 1851 |
| 16 | Brig | "Elizabeth Duncan" | 94.5' × 23.2' × 14.5' | 228 | Mr. Westland | Aberdeen | 1852 |
| 17 | Ship, Wood | "Harkaway" | 167.8' × 32.4' × 21.0' | 830 | J. & F. Somes | London | 1853 |
| 18 | " | "Polmaise" | 178.0' × 32.2' × " | 878 | Mr. Campbell | Glasgow | 1854 |
| 19 | " | "Whirlwind" | 187.0' × 33.6' × " | 1003 | J. & F. Somes | London | 1855 |
| 20 | " | "Burnah" | 185.0' × 33.6' × 21.2' | 1020 | A. Willis & Co. | Liverpool | 1856 |
| 21 | " | "Eastern Monarch" | 230.0' × 40.3' × 24.9' | 1849 | J. & F. Somes | London | 1858 |
| 22 | Barque (Wood) | "Dartmouth" | 145.3' × 34.1' × 21.6' | *978 | L. Tulloch & Co. | Sunderland | 1859 |
| 23 | Sealer & Whaler | "Ianthe" | 135.6' × 26.0' × 16.6' | 380 | Dundee S. & W. Fishing Co. | Dundee | " |
| 24 | " | "Narwhal" | 151.4' × 30.1' × 18.5' | *533 | " | " | 1861 |
| 25 | Barque (Wood) | "Star of India" | 190.4' × 34.2' × 22.1' | *1092 | J. & F. Somes | London | 1860 |
| 26 | Sealer & Whaler | "Camperdown" | 154.5' × 30.0' × 18.6' | *541 | Dundee S. & W. Fishing Co. | Dundee | " |
| 27 | Barque (Wood) | "Polynia" | 146.2' × 29.0' × 18.1' | *472 | Alex. Stephen & Sons | " | 1861 |
| 28 | " | "Earl Dalhousie" (1) | 191.5' × 34.8' × 22.2' | 1047 | Cursetjee Jamsetjee (Per Forbes & Co.) | Bombay | 1862 |
| 29 | " | "The Sir Jamsetjee Family" | 192.8' × 34.8' × 21.9' | 1049 | " | " | 1863 |
| 30 | Sealer & Whaler | "Wolf" (1) | 131.5' × 25.4' × 13.3' | 400 | Walter Grieve | Greenock | " |
| 31 | " | "Alexander" | 149.0' × 29.2' × 18.6' | 590 | Gilroy Bros. | Dundee | 1864 |
| 32 | " | "Erik" | 157.8' × 29.5' × 18.5' | *533 | G. Gibbs | London | 1865 |
| 33 | " | "Esquimaux" | 157.3' × 29.5' × 19.3' | 593 | Dundee S. & W. Fishing Co. | Dundee | " |
| 34 | Barque (Composite) | "Retriever" | 138.0' × 26.5' × 15.5' | 462 | Ridley, Son & Co. | Liverpool | " |
| 35 | " | "Corona" | 209.6' × 35.0' × 22.0' | *1202 | Alex. Stephen & Sons | Dundee | 1866 |
| 36 | Sealer & Whaler | "Nimrod" | 136.0' × 26.9' × 16.0' | *334 | Job Bros. | Liverpool | " |

*NOTE.—All Sealing and Whaling Vessels are Wood and Auxiliary Steam.*

## SHIPS BUILT BY ALEXANDER STEPHEN & SONS, AT DUNDEE—continued.

| Yard No. | Type of Ship | Name of Ship | Dimensions | Builders' Old Tonnage (Gross*) | Built for | Owners' Port | Completed |
|---|---|---|---|---|---|---|---|
| 37 | Ship (Composite) | "Sree Singapura" | 164.8' × 27.7' × 17.5' | 585 | McTaggart & Co | London | 1866 |
| 38 | Sealer & Whaler | "Mastiff" | 137.4' × 26.9' × 16.1 | *360 | John Munn | Harbour Grace | 1867 |
| 39 | " | "Arctic" (I) | 158.0' × 29.3' × 19.5' | *567 | Alex. Stephen & Sons | Newfoundland | ,, |
| 40 | — | Unnamed | Destroyed in Fire | — | — | Dundee | ,, |
| | | YARD TOTALLY DESTROYED BY FIRE | | | | | 1868 |
| 41 | Barque (Composite) | "Tonbridge" | 181.0' × 32.0' × 19.4' | 856 | J. H. Luscombe | London | 1869 |
| 42 | ,, | "Laju" | 162.6' × 28.0' × 17.6' | 556 | Dundee Shipowning Co. (W. O. Taylor, *Manager*) | Dundee | ,, |
| 43 | Steamer (Iron) | "Cheops" | 255.0' × 33.2' × 24.5' | *1505 | Alex. Stephen & Sons (Sold to Shaw Maxton and Co., 1885) | ,, | 1870 |
| 44 | Barque (Composite) | "Woodlark" | 182.4' × 32.1' × 19.3' | *800 | Alex. Stephen & Sons | Harbour Grace | ,, |
| 45 | Sealer & Whaler | "Commodore" | 151.0' × 27.1' × 16.5 | *427 | John Munn | Newfoundland | 1871 |
| 46 | Sealer | "Hector" | 151.1' × 27.1' × 16.6' | *473 | Job Bros. | Liverpool | 1870 |
| 47 | Sealer & Whaler | "Eagle" | 156.4' × 28.7' × 18.1 | *506 | N.F. Sealing & Wh. Co. (C. T. Bowring Bros.) | St. Johns Newfoundland | 1871 |
| 48 | Steamer (Iron) | "Cyphrenes" | 300.0' × 34.1' × 25.5 | *1994 | Alex. Stephen & Sons | Dundee | 1872 |
| 49 | ,, | "American" | 320.0' × 34.2' × 19.7' | 2126 | Union S.S. Co. | Southampton | 1873 |
| 50 | Sealer & Whaler | "Ranger" | 161.1' × 28.7' × 18.0 | *520 | Robert Alexander | St. Johns, N.F. | 1871 |
| 51 | ,, | "Wolf" (II) | 165.9' × 28.8' × 18.0 | *520 | Walter Grieve | Greenock | ,, |
| 52 | ,, | "Iceland" | 150.5' × 27.3' × 16.4 | *423 | D. Murray & Son | Glasgow | 1872 |
| 53 | Sealer | "Discovery" ("Bloodhound") | 160.0' × 32.2' × 18.3 | 306 | British War Vessel | London | ,, |
| 54 | ,, | "Proteus" | 190.4' × 29.9' × 18.6' | *687 | J. W. Stewart | Greenock | 1873 |
| 55 | ,, | "Neptune" | 190.5' × 29.8' × 18.1 | *684 | Job Bros. | Liverpool | 1872 |
| 56 | ,, | "Bear" | 190.4' × 29.9' × 18.6 | *680 | W. Grieve | Greenock | 1874 |
| 57 | Sailing Ship (Iron) | "Lochee" | 264.2' × 39.0' × 23.4 | *1812 | Dundee Clipper Line | Dundee | ,, |
| 58 | Sealer | "Arctic" (II) | 200.6' × 31.6' × 19.9 | *828 | Alex. Stephen & Sons | ,, | 1875 |
| 59 | Barque (Iron) | "Edith Lorn" | 200.1' × 32.3' × 19.9 | *847 | Dundee Shipowning Co. (W. O. Taylor, *Manager*) | ,, | 1876 |
| 60 | Sailing Ship (Iron) | "Duntrune" | 245.2' × 38.3' × 23.0' | *1565 | Dundee Clipper Line | ,, | 1875 |
| 61 | ,, | "Maulesden" | 245.2' × 38.3' × 23.1 | *1554 | ,, | ,, | ,, |
| 62 | Sealer | "Aurora" | 165.2' × 30.6' × 19.0 | *580 | Alex. Stephen & Sons | ,, | 1876 |
| 63 | Barque (Iron) | "Aithernie Castle" | 233.5' × 36.2' × 21.3 | *1260 | Geo. Duncan | Liverpool | ,, |
| 64 | Sailing Ship (Iron) | "Glamis" | 225.3' × 34.8' × 21.9 | *1205 | Dundee Clipper Line | Dundee | ,, |
| 65 | ,, | "Southesk" | 225.2' × 35.0' × 21.8 | *1210 | ,, | ,, | ,, |
| 66 | Barque (Iron) | "Glengarry" | 199.8' × 32.2' × 19.1 | *844 | Dundee Shipowning Co. (W. O. Taylor, *Manager*) | ,, | 1877 |

*NOTE.*—All Sealing and Whaling Vessels are Wood and Auxiliary Steam.

# Appendix A

## SHIPS BUILT BY ALEXANDER STEPHEN & SONS, AT DUNDEE—continued.

| Yard No. | Type of Ship | Name of Ship | Dimensions | Builders' Old Tonnage (Gross*) | Built for | Owners' Port | Completed |
|---|---|---|---|---|---|---|---|
| 67 | Barque (Iron) | "Stuart" | 202.5′ × 34.2′ × 19.1′ | *912 | J. Hay & Co. | Liverpool | 1877 |
| 68 | " | "Overdale" | 203.3′ × 34.2′ × 19.2′ | " | " | " | " |
| 69 | " | "Edgbaston" | 203.2′ × 34.2′ × 19.2′ | " | T. Frost, Junr. | Glasgow | 1878 |
| 70 | " | "Easterhill" | 202.5′ × 32.1′ × 18.8′ | *915 | R. Gilchrist & Co. | Dundee | 1879 |
| 71 | " | "Helenslea" (I) | 228.0′ × 35.2′ × 21.8′ | *1248 | Alex. Stephen & Sons | Paris | " |
| 72 | " | "Victorine" | 233.5′ × 36.2′ × 21.3′ | *1253 | Ant. Dom. Bordes | Dundee | 1880 |
| 73 | Sealer & Whaler | "Resolute" | 175.5′ × 30.7′ × 18.6′ | *624 | Dundee S.&W. Fishing Co. | " | 1881 |
| 74 | " | "Thetis" (I) | 181.1′ × 30.9′ × 19.1′ | *723 | Alex. Stephen & Sons (American War Vessel) | | |
| 75 | Steamer (Iron) | "North Sea" | 230.0′ × 30.9′ × 15.9′ | *1117 | Dundee, Perth & London Shipping Co. | " | " |
| 76 | Barque (Iron) | "White Sea" | 230.3′ × 30.7′ × 15.9′ | *1119 | Dundee Shipowning Co. (W. O. Taylor, Manager) | " | " |
| 77 | " | "Glenfarg" | 203.8′ × 34.1′ × 19.1′ | *898 | " | " | 1882 |
| 78 | Barque (Steel) | "Glenshee" | 203.8′ × 34.1′ × 19.1′ | *895 | Alex. Stephen & Sons | " | " |
| 79 | " | "Helenslea" (II) | 249.8′ × 35.4′ × 21.6′ | *1374 | Dundee Shipowning Co. (W. O. Taylor, Manager) | " | " |
| 80 | " | "Glenfyne" | 213.8′ × 34.2′ × 19.1′ | *957 | " | " | " |
| 81 | " | "Glenogle" | 213.8′ × 34.2′ × 19.1′ | *958 | Alex. Stephen & Sons | " | 1883 |
| 82 | " | "Earl of Dalhousie" (II) | 264.0′ × 38.7′ × 23.4′ | *1765 | " | " | 1884 |
| 83 | Steamer (Steel) | "Thane" | 245.0′ × 33.2′ × 20.6′ | *1351 | R. A. Mudie & Son | " | 1883 |
| 84 | Sealer | "Terra Nova" | 187.0′ × 31.0′ × 19.0′ | *744 | Alex. Stephen & Sons | " | 1884 |
| 85 | Barque (Steel) | "Thetis" (II) | 248.5′ × 35.4′ × 21.6′ | *1352 | " | " | 1885 |
| 86 | Barque (Steel and Iron) | "Doris" | 248.6′ × 35.3′ × 21.6′ | *1353 | " | " | 1887 |
| 87 | Barque (Steel) | "Eudora" North Carr Lightship | 287.5′ × 40.5′ × 23.7′ | *1992 | Northern Lights Commissioners | Edinburgh | 1888 |
| 88 | | | | | | | 1889 |
| 89 | Barque (Steel and Iron) | "Newfield" (I) | 248.6′ × 35.3′ × 21.6′ | 1306 | Brownelles & Co. | Liverpool | " |
| 90 | Barque (Wood) | "Diana" | 151.1′ × 24.1′ × 16.6′ | *473 | Job Bros. | " | 1891 |
| 91 | Barque (Steel and Iron) | "Galena" | 292.0′ × 42.0′ × 24.0′ | 2294 | Alex. Stephen & Sons | Dundee | 1890 |
| 92 | " | "Mayhill" | 292.0′ × 41.0′ × 23.7′ | *2121 | W. & J. Myres Sons & Co. | Liverpool | 1891 |
| 93 | " | "Annie Speer" | 243.0′ × 37.1′ × 21.6′ | 1540 | Brownelles & Co. | " | " |
| 94 | " | "Kirkhill" | 243.0′ × 37.1′ × 21.6′ | 1540 | John Steel & Son | " | 1892 |
| 95 | " | "Melita" | 310.0′ × 45.2′ × 25.2′ | 2946 | Alex. Stephen & Sons | Dundee | " |
| 96 | " | "Pitlochry" | 319.5′ × 45.2′ × 26.5′ | 3088 | Alex. Stephen & Sons (Afterwards sold to Laisz, Hamburg) | " | 1894 |
| 97 | Barque (Steel) | "Newfield" (II) | 249.2′ × 37.2′ × 21.5′ | 1512 | Brownelles & Co. | " | 1893 |

NOTE.—All Sealing and Whaling Vessels are Wood and Auxiliary Steam.

# Appendix B

## Description and Specifications of SS *Terra Nova*

The following document is taken from *British Antarctic Expedition 1910–1913 'Miscellaneous Data'*, prepared by Colonel H.G. Lyons FRS, published in 1924 and held at the Scott Polar Research Institute.

It will be seen that the footnote on the opening page, No. 17 states that the description of the ship is a pre-prepared document by Commander H.L.L. Pennell RN; Surgeon Commander E.L. Atkinson RN and Mr F.E.C. Davies, Leading Shipwright RN of the *Terra Nova*. The footnote goes on to say that 'both officers lost their lives during the war'. It should be pointed out here that Surgeon Commander Atkinson, although seriously injured aboard a warship, lived until 1929 and Leading Shipwright Davies lived beyond the Second World War. Commander Pennell died at the Battle of Jutland on 31 May 1916.

As previously stated, *Terra Nova* had no connection with the Jackson–Harmsworth Expedition of 1894–97. Reference should be made to the Fiala–Ziegler Arctic Expedition, 1903–05 (see Chapter 4).

# CHAPTER II.

## DESCRIPTION OF THE SHIP.

The *Terra Nova* was built at Dundee in 1884 for the whaling industry in the Arctic seas, and was strengthened specially on this account. Besides her whaling service, she had been in the Arctic with the Jackson-Harmsworth Expedition and in the Antarctic with the *Morning* on the relief expedition which was sent out to the *Discovery* in 1903. To prepare her for her second voyage to the Antarctic her bow and stern were reinforced with seven thicknesses of oaken beams, amounting to a total thickness of about 7 feet. Besides this there were sixteen oak beams about a foot square from the deck above the keel to the beams of the lower deck. These supporting beams and the strengthening of the bow took the strain of any impact against the ice-floes, so that the ship rose against them. The ribs, which were set close, were about a foot square in section and extended from the bow to near the stern, where six thicknesses of heavy oaken planking were used to provide additional strength. There was a 2-inch covering of steel on the lower part of the stern, but friction against the ice tended to tear this away from the woodwork.

The fitting-out of the *Terra Nova* was carried out by the Glengall Company in the East India Docks on the Thames, where they set up all the ship's rigging and renewed all the running gear. She was also sheathed with wooden skins from the bow to the break of the poop, to enable her to withstand pressure better in her passage through the ice. This sheathing extended to the upper deck and to about 4 feet below the water-line. The upper deck was in three portions: the forecastle forward, the upper deck as the middle portion, and the poop aft. At the break of the poop were the steps up to the poop deck from the main deck on either side. Both the forecastle and poop were railed. A simple swivel davit was fitted in the foremost part of the forecastle, and was useful in catting the anchors. Abaft this was a large skylight and ventilation having on either side flaps of thick glass in stout wood frames for supplying light and air to the compartment in the forecastle head; beneath this a covered iron grid provided

[NOTE.—The description of the *Terra Nova* and the discussion of the ship's journeys were undertaken by the late Commander H. L. L. Pennell, R.N., and at the outbreak of war, in 1914, he had received from Mr. F. E. Davies, Leading Shipwright R.N., the carpenter of the *Terra Nova*, the details of construction, which are printed below. Surgeon-Commander E. L. Atkinson has contributed a short general description, but as both of the officers of the ship's party lost their lives during the war it has seemed best to print this description as it stands, and to make no attempt to compile an account of the ship's journeys, which only they could have done satisfactorily.]

additional protection. Access to the forecastle from the upper deck was provided by a ladder on the starboard side. Two mooring bollards were on either side of the forecastle. Abaft the skylight was a ventilating cowl which supplied air to the sleeping quarters of the crew, which were on the lower deck below the forecastle head. Abaft the ventilating cowl was a capstan, which was used, when required, for raising the anchor and catting it.

Below the break of the forecastle on the port side were the men's W.C. and urinal, while those of the warrant officers were in a corresponding position on the starboard side, where there was also a compartment for the boatswain and his stores. Amidship on the main upper deck was a steam windlass, which was used for heaving in the anchor cable, and from here was a lead to the cable locker through a slotted hatch in the forward part of the winch.

Abaft the windlass was a hatch step leading to the crew's sleeping quarters on the lower deck. The foremast was stepped between this hatch and the galley. The galley was specially built and fitted with a naval range capable of cooking for 120 men, which was set up in the after part of the shelter.

In the forward part were shelves on which the various cooked supplies of meals for the days could be placed, to be removed by the watches as they were required. The galley was very strongly built, and was securely bolted through the upper deck; it was provided with doors on either side so that the lee door could be left open for ventilation. Situated conveniently near the crew's mess, its distance from that of the officers was an inconvenience in rough weather.

On either side forward, below the break of the forecastle, were two skids of stout timber and on them two clinker-built double-ended lifeboats lay in their chocks. The position was a good one, and covered with canvas covers and secured by seizing ropes and grapplings, they withstood the worst weather. Abaft the galley was an accessory steam winch which was rarely used.

The ice-house which was erected abaft of it was tin-lined and had four thicknesses of wood besides thick felt. This was apparently insufficient to maintain a constant temperature for some of the meat stored in it went bad.

On the top of the ice-house were mounted the standard compass, a Lloyd-Creak instrument and a rangefinder for use when making running surveys. Abaft of the house was the main hatch, and next came the main mast and on either side of this a Stephens pump, while amidships at the break of the poop-deck was the small after hatch.

The forward part of the poop was occupied by the funnel, and forward of it was a steam winch, while between it and the funnel were two iron tanks for thawing ice when this could be obtained. On either side of the after portion of the funnel were gratings and the two cowls and ventilating shafts leading to the stokehole. The remainder of the mid-ship portion was occupied by a strong deck-house forming a protection for the enlarged wardroom, and an alleyway above this. The forward portion of this deck-house was used by the navigator as a chart-room.

On the starboard were two w.c.'s for officers and in-board from them were the steps leading to the bridge, the mid-ship portion of which was occupied by a wooden shelter in which was a large shelf for charts. A few steps ran from the mid-ship portion of the bridge to others which led to the top of the deck-house on which was a compass. Astern of the deck-house was the compass, and the wheel, which was double, and was fitted with a foot-brake. On either side two grills occupied the greater portion of the deck, and above the stern on the out-board side of these were two mooring bollards, port and starboard.

On the out-board side, port and starboard, there were two whaling boats slung from davits.

Up to the time of our departure from New Zealand for the Antarctic the forecastle head was used for messing accommodation by the crew, but then it was required for the ponies until they were landed at Ross Island in the Antarctic. It was sufficient for fifteen ponies and four stalls for the others were erected between the ice-house and the galley.

In the second year stalls were built on the upper deck for the mules, but this would have been impracticable in the first year. The accommodation for the officers and scientific staff consisted of a wardroom 24 feet long and 9 feet wide, down the centre of the ship, which was lighted by skylights in the top of the deck-house and by small ports at the sides of the cabins which opened off it. On the port side there was a large cabin with six bunks, and two smaller ones containing three and two bunks. On the starboard side there was a large cabin forward, and two smaller cabins each with two bunks, the remainder of the space being taken up by the alley-way and the steps leading to the upper deck. Over the stern there was a cabin on either side, each containing three bunks.

In the forward part of the wardroom was a stove, and forward on the port side was the entrance door to the pantry, which was fitted with cupboards, etc., for china, glass and cutlery.

Beneath the wardroom was the lazarette which was entered by a hatch in the alley-way forward of the wardroom entrance. In it was stored the bulk of the expendable stores, food, etc. Immediately below this was the tunnel for the propeller shaft. Forward of the lazarette was a small compartment which was used as a chronometer-room and for the storage of other delicate instruments. Forward of this from the keel to the upper deck was the engine-room and boiler-room.

The engines were vertical compound engines working at an initial pressure of 60 lbs. to the square inch, and with them the maximum number of revolutions per minute for going ahead was 60. After the first year alterations were proposed by Chief Engine-Room Artificer W. Williams to obtain increased efficiency, and these were carried out by M. Dickson of Lyttelton, New Zealand, with the result that they could then develop 89 revolutions per minute when going ahead and 60 when going astern.

The propeller shaft was exceptionally thick and of the best steel, and the worth of this was appreciated when, as not infrequently happened, the engines were brought to a dead stop by masses of ice jamming the propeller.

The necessity for the bedding of the boilers was questioned at one time, but they stood many trials, although the ship under some conditions rolled very heavily, as much as 50° each way, and was practically beam-ended on at least four occasions.

Forward of the engine-room bulkhead the lower deck space was divided between coal and general stores. Coal was stowed loose in the after third, and it could, by means of a hole in the bulkhead, be trimmed into two small bunkers on either side of the stokehold. A wooden bulkhead separated the coal from the general stores which were stowed in the forward two-thirds of the upper main hold.

Fresh water was stored in tanks holding 8·15 tons each, which were on the lower deck abaft the foremast. Two out of the four were used for this purpose, the other two being filled with compressed fodder for the animals. All the lower hold abaft these tanks was filled with coal.

The *Terra Nova* was three-masted, the fore and mainmasts being stepped above the keel, but the mizzenmast was stepped into a reinforced portion of the lower deck in the forward portion of the wardroom pantry.

The following details of the ship have been supplied by Mr. F. E. Davies, Leading Shipwright R.N., and those of the engine-room by Mr. W. Williams, Artificer Engineer R.N., while Messrs. David Bruce and Co., of Billiter Square Buildings, E.C., kindly sent prints of the plans of the ship.

S.Y. *Terra Nova*.

Built in 1884 at Dundee.

Tonnage register, 399·7 ; displacement, 858 tons (light).

| | |
|---|---|
| Length | 187 feet (over all). |
| Breadth | 31 feet (extreme). |
| Depth of hold | 19 feet. |

Barque rigged ; royal yards not carried this commission.

PARTICULARS OF CONSTRUCTION.

*Keel.*—Of American rock elm, 14¾ inches by 14¾ inches ; scarphed together after French style. Length of scarph, 6 feet 3 inches ; lip, 3¼ inches.

*Frames.*—Of German oak sided to 13 inches or 14 inches. Maximum depth of frame, 12¾ inches, tapering to 6¾ inches at covering board ; breadth of frame, 12¾ inches.

*Keelson.*—Of pitch pine, 18 inches by 16 inches, running fore and aft from stemson to sternpost, tapering at fore and after ends. Scarphs of keelson, 7 feet 3 inches, with 4½-inch lip. Bolted through keel at every alternate timber by 1¾-inch bolts.

*Sister keelson.*—One each side of keelson, 2 feet 8 inches from it ; 14 inches in width and trimmed to shape of frames ; secured through each alternate timber by 1½-inch bolts.

*Planking of bottom.*—(Outer) of Canadian elm or American rock elm, from keel under bottom and on turn of bilge to a height of 7 feet 7 inches from base line. Sides of pitch pine. *Planks* are 1 foot by 4 inches in bottom, garboards being slightly thicker. From bottom of doubling to 7 feet 7 inches from base line they increase in thickness gradually from 4 inches to $5\frac{1}{4}$ inches. Secured by $\frac{3}{4}$-inch bolts at butts and by hardwood trenails ($1\frac{3}{8}$-inch) at frames. Side planks 1 foot by $5\frac{1}{4}$ inches up to within five planks below covering board; from there they gradually taper in thickness until it is reduced to $4\frac{1}{4}$ inches at covering board.

*Doubling.*—Extends fore and aft from 3 feet above base line to 18 feet 6 inches above base line, of iron bark $2\frac{1}{4}$ inches thick, except two top strakes which are 2 inches and bottom strake $2\frac{1}{2}$ inches thick. Fastened by 11-inch and 9-inch bolts.

*Ceiling in bottom.*—Pitch pine, $5\frac{1}{2}$ inches thick in bottom and $4\frac{1}{4}$ inches at ship's side.

### Main Deck.

*Shelf.*—$13\frac{1}{4}$ inches thick at top, tapering to $6\frac{1}{2}$ inches and 18 inches deep.

*Main Deck Beams.*—Of pitch pine ($12\frac{1}{2}$ inches by $12\frac{1}{2}$ inches); joggled 2 inches over shelf and secured by hanging and lodging knees with seven bolts in each. Beams about 5 feet 3 inches apart from centre to centre. Main deck stringer same size as shelf.

### Upper Deck.

*Upper Deck Shelf.*—$10\frac{1}{4}$ inches thick on top, tapering to 5 inches and 15 inches deep.

*Upper Deck Beams.*—$9\frac{1}{2}$ inches by 10 inches and about 5 feet 3 inches apart centre to centre; stringer, 12 inches by 12 inches.

*Covering Board.*—4 inches thick.

### Quarter Deck (Poop).

*Upper Deck.*—Stringer and shelf carried eight frame spaces under quarter deck.

*Shelf.*—Under quarter deck 5 inches at top, 3 inches at bottom, 12 inches deep.

*Beams.*—Of German oak ($8\frac{1}{2}$ inches by 9 inches), again 5 feet 3 inches apart.

*Stringer.*—9 inches by 9 inches.

*Decks.*—Upper Deck and Poop of pitch pine, 6 inches by 3 inches. Top gallant forecastle Deck of yellow pine, 6 inches by 3 inches.

*Hatch Coamings.*—Pitch pine. Main hatch, $14\frac{1}{2}$ inches by 5 inches. Fore and after hatches, 8 inches by 6 inches. Top of coaming, 1 foot 5 inches above line of beam.

*Hatches.*—$2\frac{1}{2}$ inches thick.

*Iron Beams* under top gallant forecastle, 5 inches by 3 inches by $\frac{8}{15}$ inch, 4 feet 6 inches from centre to centre.

*Bulwarks.*—Stanchions, $7\frac{1}{4}$ inches by $6\frac{3}{4}$ inches at bottom; $7\frac{1}{4}$ inches by 6 icnhes at **top**. Bulwarks of pine $1\frac{1}{4}$ inches thick, top and bottom strake being $1\frac{1}{2}$ inches.

The false stem, stem, stemson and deadwood of English oak, totalling to about 6 feet through at the water line.

Iron sheathing for protection against ice extended from the bows 6 feet aft and to 7 feet above and below the normal water line. She is also fitted with a heavy iron stem band.

On the inside of the bows below the main deck from the stem to 45 feet aft two extra tiers of beams were worked immediately below the main deck beams and well pilloried from the keelson. Also for this distance riders (12 inches by 12 inches) were worked inside the frames, diagonally across and 1 foot apart.

As originally built she had a screw well for lifting the screw but this had been filled in when the large propeller was fitted.

The rudder was plated with ⅝ inch mild steel plates to about 6 feet below the normal water line. The weight of the wood in the rudder is nearly 1 ton and of the iron plating another ¾ ton. It is hung by 4 pintles and originally there was nothing to prevent the rudder unshipping except its weight. After the first season (1910–11) a wood lock was fitted for extra safety in case of a lifting force from pressure.

### Masts and Yards.

#### Main Mast.

|  | feet | inches |  | feet |
|---|---|---|---|---|
| Lower Mast | 70 | 6 | Housing | 18 |
| Topmast | 37 | 0 | ,, | 9 |
| T'gallant mast | 36 | 6 | ,, | 6 |

Height of Truck above U.D., 111 feet.

#### Fore Mast.

|  | feet | inches |  | feet |
|---|---|---|---|---|
| Lower Mast | 68 | 0 | Housing | 18 |
| Topmast | 37 | 0 | ,, | 9 |
| T'gallant mast | 36 | 6 | ,, | 6 |

Height of Truck above U.D., 108 feet 6 inches.

#### Mizzen Mast.

|  | feet | inches |  | feet | inches |
|---|---|---|---|---|---|
| Lower Mast | 58 | 0 | Housing | 7 | 0 |
| Topmast | 48 | 0 | ,, | 7 | 0 |

Height of Truck above Poop, 92 feet.

## Main and Fore.

|  | Length. feet inches | Diameter. Bunt inches | Diameter. Quarter inches |
|---|---|---|---|
| Lower Yard | 64  0 | 14 | 13 |
| Upper Topsail Yard | 53  0 | $12\frac{1}{4}$ | 11 |
| T'gallant Yard | 41  0 | 0 | 9 |

## Bowsprit.

|  | feet inches |  | feet inches |
|---|---|---|---|
| Length | 49  3 | Housing | 16  6 |

Total length outboard, 32 feet 9 inches. Tapering from $16\frac{1}{2}$ inches diameter at heel to 8 inches.

### General Remarks.

*Coal Space.*—By original design 130 tons could be carried on the main deck forward of the boiler room bulkhead. The tanks being removed from the hold this was used for coal, and then the hold from boiler room bulkhead to the fresh water tanks up to the main deck, and above the main deck from the boiler room bulkhead to the fore end of the main hatch, together with the 68 tons in the bunkers, brought her coal capacity up to 550 tons (New Zealand coal).

From the fore end of the main hatch to the foremast was then available for general cargo and extra living space.

*Water Tanks.*—In the hold, 10 feet abaft foremast and height to the main deck beams; four in number. Two rectangular ones, held $12\frac{1}{2}$ tons each, and two wing tanks fitting the shape of the ship held 8 tons each.

Though in the design counted as ballast tanks they were all used, when required, for fresh water.

*Crew Space.*—The topgallant forecastle was used as a mess deck and the forecastle as a sleeping deck. The men slept in hammocks, all bunks being removed from both forecastles, and lockers were arranged along the ship's side for kits. On the port side of the lower forecastle abreast the foremast a warrant officers' mess was built with six bunks.

An instrument room of match boarding was built starboard side lower forecastle, opposite the warrant officers' mess to hold all instruments. It was, however, found to get very wet and no delicate instruments could be stowed there.

The wardroom was much enlarged and, including the pantry, extended from the engine-room bulkhead aft. Cabins were arranged all round, and the original engineers' mess now opened into the wardroom and was fitted up as a six-bunk cabin. Total number of bunks, 24.

The skylight and companion to the wardroom were removed, and a deck house built over the wardroom for light and air. The foremost end of this house round the mizzen mast was made into a chart-house, and it also allowed of a gangway about 2 feet 6 inches broad on the starboard side, between the after door and the companion down to the wardroom (the old hatch to lazarette). This house was constructed of two thicknesses of 1 inch spruce boards.

The gangway mentioned above was found almost indispensable in cold weather for the biologist, the baths into which the contents of a trawl were emptied, immediately it was got in-board, being at once put in here to prevent the catch freezing, and by keeping the wardroom fire up it was possible to keep the temperature here slightly above freezing point even with 20 to 30 degrees of frost outside.

On the port side of the poop four laboratories for scientific work were built of 1 inch spruce double thickness. The after one was fitted up as a photographer's dark room. The foremost one had two clome troughs with small tanks overhead (one salt water and one fresh water) for the use of the biologists, while all were fitted with shelves for bottles, etc. By exercising care in replacing bottles there was hardly a breakage in spite of the excessive motion which was sometimes experienced. These houses were found to be unsatisfactory as they soon leaked badly through the expansion and contraction of the wood, causing great annoyance to the scientists and damage to their gear. It would be advisable on any similar occasion to use teak in spite of the greater expense.

The compartment in the hold before the sail-lockers was used as a biological locker, it being about 8 feet fore and aft and the breadth of the ship.

As a contributory cause towards the freedom from loss due to breakage must be counted the fact that, as soon as possible after preserving, the bottles containing biological specimens were safely packed in wooden boxes and stowed in this locker. Water samples, however, which were stowed here froze and broke the bottles.

On the upper deck between the main and fore hatches an ice house was constructed of:—outside, two thicknesses of 1 inch spruce boarding grooved and tongued; lagged top, bottom and sides, with a non-conducting material. Inside were another two thicknesses of 1 inch spruce boarding, and the bottom and 3 feet up the sides lined with lead. There were two drains to let the melted ice water run off. This house when filled with ice and meat was such a heavy weight (about 10 tons) that its position on the upper deck was a source of weakness. On top of the ice house were placed the Standard Compass, the Dip Circle gimbal stand and the Range Finder.

Below the wardroom over the shaft alley was a lazarette extending from the engine room bulkhead to the body post (22 feet), which was used for stowing " present use " stores in. Down here was also the chronometer room. The lazarette was the only really permanently dry place in the ship.

There were no watertight bulkheads throughout the ship.

Two leaks were a source of considerable trouble till they were finally located. One of the through bolts in the stem had become slack in the hole allowing free passage for water, but there were also other leaks about the bows beneath the sheathing; the original caulking in the ship's planking beneath the sheathing having become perished from age, and the ship having had such frequent small repairs, it was difficult without a large refit to make the sheathing itself watertight.

In the stern, by careless workmanship at some time, a hole in the ship's planking beneath the sheathing had been left unplugged near the screw well. Water thus found its way easily inside the sheathing when it had a free passage inboard. This leak was very difficult to locate and was not found for nine months.

At her worst, when fully laden, she made about 1 foot 6 inches of water in a watch, and when these two leaks were stopped it was reduced to this amount per day, but after working in the pack there was always an increase of leak forward which could only be stopped properly by stripping down the sheathing in the bows and recaulking the hull.

*Hand Pump.*—Abaft the mainmast, four 6-inch plungers (two to each suction pipe) about 9 inch stroke.

The suction pipes as originally fitted were heavy cast-iron pipes in two lengths (*i.e.*, each length about 10 feet) running down between the frames and open at the bottom with no rose fitted.

A trunk gave access to the well from the after hatch, but the space in which the bottom of the suction pipe was, was so small that it was always difficult to clear it of coal even when it was practically dry, and almost impossible to do so at times when there was 3 feet of water in her.

After the first year's experience the lower end of the suction pipe was cut and flanged 3 feet 6 inches from the bottom, and a light iron pipe and rose fitted which could easily be removed if required.

In bad weather it was impossible to open the after hatch and so get down to the pump well, while from forward the way was blocked by coal and stores (when fully laden). It was necessary therefore in the gale of December 8, 1910, to cut a man-hole in the boiler-room bulkhead to allow access to the trunk.

After the first season the bilges were thoroughly cleaned and the ceiling, which was damaged, repaired. The bilges had become filled up with coal that had found its way through the damaged ceiling and mixed with the residue oil from the engine room bilge. This mixture formed into large balls that soon choked the pumps.

The original pump handles were short handles allowing of only three men aside, but these were replaced at Simonstown by long portable crank handles between the pump and ship's side on both sides of the ship.

Without these better handles it would often have been impossible to pump out in the Southern Ocean.

*Skids.*—Immediately abaft the foremast three skid beams were built the height of T'gallant forecastle. On these were stowed a whaler and cutter, and also the ship's spare timber for repairs. These skids were found of the greatest use.

## Stables.

### First Year—19 Ponies.

The top gallant forecastle was fitted up into fifteen stalls, the lamp room being removed.

Each stall was 2 feet 8 inches broad with portable partitions between. The front boards were fixed, but this was found to be a great disadvantage if an animal fell.

A trunk was built round the hatch leading from the top gallant to lower forecastle and carried up to the skylight so that the air the men got in the sleeping deck below should come direct and not through the stables.

Four stalls were built on the port side of the fore hatch, between the ice house and the after skid beam, facing inboard. These were strongly constructed of 3 inch deals, bolted through the deck, receiving some support from the ice house and after skid beam, and also secured by chains to deck ringbolts. The partitions were of 2-inch red pine.

### Second Year—Seven Mules.

The mules were all carried outside.

The stables were again built round the fore hatch, four stalls being on the port side and three on the starboard, all facing inboard. The deck between being brought up to the level of the hatch coaming.

The front boards were fitted to be easily removed to take out any one animal at a time. (See sketch below.)

At the bottom (in front) a wide board close down to the deck prevented the possibility of their feet slipping under.

The partitions were again removable but brought close down to the deck for the same reason.

On the floor of the stalls cocoanut matting was put down and 1 inch × 2 inch battens, to give the animals a hold. The matting choked the drain holes and was found to get very foul, and was therefore discarded after a few days.

Eye-bolts were fitted on the top of the stalls to assist in lifting the animals if they should fall.

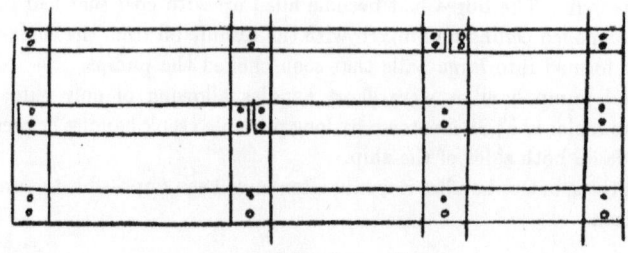

Fig. 1.

The stables were covered with painted canvas, and an awning that could be drawn across from stable to stable in bad weather was fitted.

*Dogs.*—Going south there was not room to place special fittings for the dogs, but for the passage home wooden trays were made for them to lie on. These trays were about 18 inches to 2 feet broad with a 2 inch batten running along the outer edge for the dogs to press their feet against when the ship was rolling.

They were raised 3 or 4 inches off the deck and had holes bored through to allow of free passage of air. In both hot and wet weather they were the greatest comfort to the dogs.

### Engine-Room Particulars.

The propelling machinery consists of :—

One set of 2 cylinder (H.P. and L.P.) compound vertical inverted engines made by Messrs. Gourley, of Dundee, in 1885.

| | |
|---|---|
| Nominal horse-power | 140 |
| Diameter of cylinders—H.P. | 27 inches. |
| L.P. | 54 ,, |
| Length of stroke | 2 feet 9 inches. |
| Diameter of shafting | $10\frac{1}{2}$ inches. |

The H.P. and L.P. slide valves were flat double ported.

One surface condenser was fitted, the circulating water passing through the tubes.

The air pump (1 double acting) was worked direct off the main engines.

The thrust block was of solid block type having six collars on the shaft, circulating water passing through the top cap.

The starting and reversing gear was of the " all round " description.

The main bearings, crankhead and eccentrics were lubricated by means of worsted syphons.

The propeller was four-bladed and of cast steel.

### Engine Room Pumps.

*Circulating Pump*, worked direct off the main engines, for circulating water through the condenser, could if required be put on to engine-room bilge suction.

Capacity about 600 tons per hour.

*Mail Bilge Pumps.*—Two worked direct off the main engines. Diameter of plungers 4 inch, stroke 2 feet.

*Feed Pumps.*—Two worked direct off main engines. Diameter of plungers 4 inch, stroke 2 feet.

*Evaporator.*—Makers " Kirkaldy." Output 6 tons per day. The feed pump for evaporator was worked direct off the main engines taking its suction from main condenser discharge. Diameter 2 inch, stroke 10 inches.

*Injector.*—One; for auxiliary feed and make-up feed. Makers, "Davies Metcalfe."
One auxiliary fire and bilge pump (independent of the main engines) made by "Gourley" with connections to fire main, engine room and stokehold bilges.

*Boiler.*—One marine return-tube boiler, having three furnaces. Working pressure, 80 lb. per square inch. Dimensions: diameter, 15 feet; length, 12 feet. Weight of water at working height, 26 tons.

*Ventilation.*—All natural draught. Two cowls at after end of engine room as uptakes. Two cowls over boiler room as down-takes, exhausting through the fire and up-funnel.

The space over the boiler was also ventilated by "ports" on top of the boiler room casing.

There was no bulkhead between the engine room and boiler. In the tropics the boiler-room was extremely hot.

The bunkers were ventilated by port-holes in the forward end and bunker lids on the poop.

The ship was fitted with two steam winches; two cylinder, reversing type; steam taken from boiler and exhausting to atmosphere.

The foremost winch, situated abaft the galley, had

Cylinder diameter, 8 inches.
Stroke, 12 inches.

It was used for trawling, and for working cable by means of a messenger chain and sprocket wheels on the cable holder. In cold weather the messenger chain became too brittle for this strain and snapped.

The after winch, situated at fore end of poop, had

Cylinder diameter, 6 inches.
Stroke, 12 inches.

It was used for trawling, and had a messenger connection to work the hand pump, but this was never used owing to the waste of steam.

Heaving in a trawl caused a loss of water varying from $\frac{1}{2}$ ton to 1 ton.

### REMARKS.

The boiler was examined and drill tested in 1910 and new plain tubes were fitted. The stay tubes were passed as good for the commission; they were supposed to be the original stay tubes fitted.

The centre combustion chamber was found to be only 3/16ths of an inch instead of $\frac{5}{8}$ in one part. This had already been patched unsatisfactorily, but the thickness was increased to 5/16ths by the acetylene welding process, which proved quite satisfactory.

Only fresh and distilled water was used in the boiler the whole commission.

Corrosion was found taking place on the stern shaft near the propeller, so a zinc protector was made and secured to the shaft. The propeller was cement-washed and no more action occurred.

For melting ice two iron tanks were fitted one on each side of the boiler-room casing, with a steam jet entering near the bottom and discharges leading to the boiler and ship's tanks. A little ice being melted in them, and the water raised to boiling point, it was found that ice was melted in them almost as quickly as it could be put in. As first fitted the steam was kept enclosed in a spiral tube at the bottom, but after experience this tube was removed and the steam let direct into the tank.

While crossing the Atlantic in 1910, trouble was experienced owing to hot bearings. It was found that the main bearings had been lined up with tin liners under the different brasses and these liners had deteriorated. After brass liners had been fitted in New Zealand there was no more trouble from this cause.

*Coal.*—She took in 450 tons Crown Patent Fuel at Cardiff in June, 1910.

The bricks were two sizes, of 25 lbs. and 12 lbs. weight respectively. This fuel was very suitable for stowage (35 cubic feet per ton). It was easy to trim from the hold into the bunkers and was also convenient for use with a shore party owing to its compact stowage and easy handling. It was not found to deteriorate during two winters. For use in the ship it was found not so economical as coal proper, owing to the comparatively large percentage of ash and clinker.

From New Zealand both Westport and Blackball coal were found to give good results and to be economical. They were used mixed, and worked well, the percentage of ash being only about 3 per cent.

The Blackball coal requires good bunker ventilation, and great care had to be taken that it should be received on board perfectly dry.

To ventilate the coal in the hold six wooden trunks were fitted from the bottom of the hold to above the coal, and these were perforated with holes at frequent intervals. They also allowed the temperature of the coal to be taken daily. Whenever the weather permitted the hatches were uncovered during the daytime. On two occasions the coal heated sufficiently to cause some anxiety.

*Oil.*—The lubricating oil used was No. 1 mineral engine oil of the Vacuum Oil Company and was quite satisfactory.

# Appendix C

## Extract from the Log of USCGC *Atak*

Page 3

        1830 Moored to dock at BW1 outside of NEVADA.
        1840 REINHARDT, John (234-039) B.M.2c. and LYSAGHT, Marshall
        C.B.M.(R) (582-052) reported aboard from IKARRA with ships
        equipment.

Thursday 9 September, 1943.

    Position        0800            1200            2000

    Lat.            61 08N          61 08N          61 08N
    Long.           45 25W          45 25W          45 25W

    0000 to 2400
        Moored starboard side to NEVADA at Narsarssuak, Greenland.
        1056 Underway to shift berth
        1105 moored to end of dock portside to.
        1410 Arvek moored to starboard side.

Friday 10 September, 1943.

    Position        0800            1200            2000

    Lat.            61 08N          61 08N          61 08N
    Long.           45 25W          45 25W          45 25W

    0000 to 2400
        Moored portside to dock, with Arvek to starboard at Bluie West
        One, Greenland.

Saturday 11 September, 1943.

    Position        0800            1200            2000

    Lat.            61 08N          In              60 47N
    Long.           45 25W          Skovfjork       46 18W

    0000 to 2400
        Moored portside to end of dock, B.W.1, Greenland with Arvek
        to Starborad.
        0834 Pursuant to SOPA, Greenland dispatch 102036 Sept. 1943.
        underway proceeding to Simiutak, Greenland.
        0845 Barge moored to starboard side for transportation of Simiutak.
        1800 Anchored in Gate's Hole with 30 fathoms of chain in 10 fathoms
        of water with starboard anchor. Beacon 1, 76°; Beadon 2, 315°;
        Beacon 3, 119°.

Sunday 12 September, 1943.

    Position        0800            1200            2000

    Lat.            60 57N          60 47N          Off Simiutak Island.
    Lon.            46 18W          46 18W

    0000 to 2400
        Anchored in Gate's Hole, Greenland to Starboard anchor with 30
        fathoms of chain in 10 fathoms of water.
        0605 Weighed anchor and got underway. 0715 Reversed course and
        headed back to Gate's Hole  0756 Anchored with 45 fathoms of
        Chain to starboard anchor in 20 fathoms of water in Gates Hole.

0900 Commenced hoisting cargo aboard
1215 Completed hoisting cargo aboard.
1350 Underway for Simiutak Island with barge in tow.
1745 Anchored with 45 fathoms of chain to starboard anchor in 20 fathoms of water at Simiutak Island.
1920 Received SOS distress call from TERRA NOVA. Anchorge barge with load of three dories using ship's port anchor and 55 fathoms of 2" manila line.
1930 Underway out of Simiutak Anchorage proceeding to point 60 15N, 46 48W to aid TERRA NOVA reported indistress on about 2800 Kcs.
1940 Streamed log, reading 00.0 abeam of Kioki Rock.
   Set course at 194° at 280 RPM.
2200 Changed course to 200°T.
2225 Sighted TERRA NOVA Off Port Bow.
2234 Changed Course to 104° at 180 RPM to head for TERRA NOVA.

Monday 13 September, 1943.

| Position | 0800 | 1200 | 2000 |
|---|---|---|---|
| Lat. | 60 30N | 60 45N | 61 08N |
| Long. | 46 30W | 46 23W | 45 25W |

0000 to 2400
Standing by SS TERRA NOVA. Working engines at slow speed and circling ship in distress. 0115 Master of TERRA NOVA spoke of intentions of abandoning ship and asked for ATAK to come alongside and rescue crew. 0120 Pumping oil over side to decrease sea. 0215 Lifeboat from TERRA NOVA alongside with four men. Received information from these men that 20 men were left on board TERRA NOVA. Standing by to remove rest of crew at daybreak.
0445 commenced removing crew. Boat sent over four times to pick up rest of crew. 0600 Master of TERRA NOVA aboard from last boat load. 0550 TERRA NOVA burst into flames.
0630 commenced firing 3"23 into burning derelict.
0655 Ceased firing. 22 rounds expended. Derelict sinking slowly. 0657 proceeded toward USS LAUREL sighted on horizon, lifeboat of TERRA NOVA being towed astern. 0705 standing by USS LAUREL. 0723 tow line of boat parted and boat adrift.
0725 set course at 335° for Narsarssuak at 280 RPM.
1000 Changed course to 325°. 1005 Changed course to 310°.
1612 Moored portside across end of dock at Narsarssuak.

Tuesday 14 September, 1943.

| Position | 0800 | 1200 | 2000 |
|---|---|---|---|
| Lat. | 61 08N | 61 08N | 61 08N |
| Long. | 46 26W | 46 26W | 46 26W |

0000 to 2400
Moored as before starboard side to dock outboard of IZARRA.
FAUNCE AND MANITOU at Narsarssuak, Greenland.

# Appendix D

## List of Captains Who Commanded the SS *Terra Nova*, 1884–1943

(Note: sometimes whaling ships sailed with two master mariners, the second acting as navigating officer.)

| | | |
|---|---|---|
| 1884–88 | Captain Alexander Fairweather | Dundee |
| 1888–93 | Captain Charles Dawe | St John's\Dundee |
| | Captain W. Archer | |
| 1894–97 | Captain Harry McKay | Dundee |
| 1898–1903 | Captain Arthur Jackman | St John's |
| 1903–04 | Antarctic relief\Captain Harry McKay | Dundee |
| 1905 | Arctic relief\Captain Johan Kjeldsen | Tromsø, Norway |
| 1906–08 | Captain Abram Kean | St John's |
| 1909 | Captain Edward Bishop | St John's |
| 1910–13 | Antarctic Expedition\ | |
| | Lieutenant E.R.G.R. Evans\Lieutenant H.L.L. Pennell | |
| 1913 | Captain James Paton | Rock Ferry, Liverpool |
| 1913–15 | Captain William J. Bartlett | St John's |
| 1916 | Captain Samuel R. Winsor | St John's |
| 1917–19 | Captain Nicholas J. Kennedy | St John's |
| 1920–26 | Captain Abram Kean | St John's |
| 1927 | Captain Wes. Kean | St John's |
| 1928 | Captain John Parsons | St John's |
| 1929 | Captain Jacob Kean | St John's |
| 1930–31 | Captain John Parsons | St John's |
| 1932–33 | Captain Abram Kean | St John's |
| 1934–40 | Captain Stanley Barbour | St John's |
| 1940 | Captain B. Richard Badcock | St John's |
| 1941 | Captain Stanley Barbour | St John's |
| 1942–43 | On charter\Newfoundland Base Contractors Ltd | St John's |
| | Captain Llewellyn Lush | |

## Biographies of Captains of SS Terra Nova

*Captain Alexander Fairweather (1853–96)*   *(Terra Nova 1884–88)*
Alexander Fairweather was born at Dundee in 1853 and at the age of 10 first went to sea as a cabin boy on a coaster. He was a harpooner on SS *Camperdown* in 1866 and in 1887 was appointed mate on the whaler, *Victor*. At 26 years of age he was captain of the yacht *Diana* and then commanded the whaler *Active*. His brother, James was a whaling master. In 1892, under his command in the whaler *Balaena*, together with three other whalers, *Active*, *Diana* and *Polar Star*, he led the Dundee Antarctic Whaling Expedition, 1892–93, to investigate the potential of whaling and seal fisheries in that region. He died in 1896.

*Captain Charles Dawe (1845–1908)*   *(Terra Nova 1888–93)*
Charles Dawe was born on 28 February, 1845, at Port de Grave, Conception Bay, Newfoundland, and was involved in sealing and fishing from an early age. In 1874, he had his first command of a sealer and over a twenty-two-year period he commanded the *Aurora, Bear, Iceland, Newfoundland, Thetis* and *Terra Nova*. He was successful in the Labrador cod industry and was also a politician. He died on 30 March 1908.

*Captain Harry McKay (1857–1925)*   *(Terra Nova 1894–97 & 1903–04)*
Henry Duncan McKay was born at Dundee in 1857 and gained his master's ticket in 1882 and the following year took over as master of the whaler *Aurora*. When whaling in Baffin Bay in 1893, he discovered the wreck of the *Ripple*, a Swedish survey vessel which had been wrecked by ice. He discovered body remains, papers and some personal possessions which indicated the work of the scientists Björling and Kallstenius whose ill-fated expedition had aimed to be the first to the North Pole. For this work, Captain McKay was awarded a Swedish Anthropologists Medal. A company record shows him to have commanded the *Terra Nova* from 1894 to 1897. He was appointed by the Admiralty to command the *Terra Nova* for the Antarctic relief voyage 1903–04. He retired from the sea in 1909 and lived at Tayport where he died on 9 November 1925. A cape on the south side of Ross Island, Antarctica, is named after him.

*Captain Arthur Jackman (1843–1907)*   *(Terra Nova 1898–1903)*
Arthur Jackman was born in 1843 and was 22 when he took his first vessel to

the ice. 'Viking' Arthur, as he was known, commanded the *Hawk, Falcon, Narwel, Resolute, Eagle II, Aurora* and *Terra Nova* and reputedly never lost a season at the ice. In 1881 when returning from Scotland, the funnel of the *Resolute* was washed away in a storm. He had his men construct a makeshift funnel out of sheet iron, packing cases and tin cans and continued to Newfoundland. He commanded Robert Peary's ship on his first Arctic expedition to Greenland in 1886 and in 1906 was fined for killing seals on a Sunday, in violation of sealing laws. He became Marine Superintendent for Bowring Brothers Ltd. He died on 31 January 1907.

*Captain Johan Kjeld Kjeldsen (1840–1909)* (Terra Nova *1905*)
Born at Bakkejord, Tromsøysund, Norway on 27 July 1840, the son of a blacksmith. He went to sea as a boy and by 1856 had sailed to the Arctic polar region and subsequently gained his qualifications to captain. He was a sealer, ice pilot and master of many ships and took part in a number of scientific expeditions in the Arctic region. In 1905 he was in command of the *Terra Nova* and relieved the Fiala–Ziegler Arctic Expedition. He died in 1909, after a distinguished career at sea.
(Norsk Polarinstitutt, Tromsø)

*Edward Ratcliffe Garth Russell Evans RN (1881–1957)* (Terra Nova *1910 & 1913*)
'Teddy' Evans was born in north-west London where he spent his early childhood. After attending the Merchant Taylors School, he was admitted to the mercantile marine training ship, *Worcester*, from where he gained an officer cadetship in HMS *Britannia*, Dartmouth. He was a sub-lieutenant in the *Morning* for the relief of the *Discovery* in 1903 and 1904. He joined the *Terra Nova* Expedition in 1909 as Scott's deputy leader and captain of the ship. He was invalided home from that expedition in March 1912, and after recovering returned to take command on Scott's death. He resumed his career in the Royal Navy through the First World War and served in a number of naval appointments until his retirement as Admiral of the Nore in 1938. During the Second World War, he was appointed as one of two Regional Commissioners for London with responsibility for the civil population. He entered the House of Lords in 1945 and took the title Baron Lord Mountevans of Chelsea. He died in Norway at the family retreat in 1957.

*Harry Lewin Lee Pennell RN (1882–1916)*  (Terra Nova *1911–13*)
Harry Pennell was born at Kingsthorpe in 1882. He was one of five children, his father was an army colonel serving in India. He entered the Royal Navy at Dartmouth in 1898 and in 1899 was awarded the Admiralty Study Subjects prize, an inscribed brass and copper four-face clock, with compass, barometer and thermostat which remains in the family. Harry Pennell was a midshipman in HMS *Goliath* and served in the Far East. He was commissioned as a lieutenant in 1903 and was navigating officer in the *Terra Nova* for the passage to the Antarctic. He took over the ship as captain for the intermediate voyages to relief the expedition from New Zealand. He was highly thought of and considered by many to be the most competent officer aboard the ship. Following the expedition, he resumed his naval career and was promoted to commander. He entered the war and died while serving as navigating officer in the battle cruiser HMS *Queen Mary*, under the command of Captain Cecil I. Prowse, which was sunk by enemy gunfire at the Battle of Jutland on 31 May 1916.


*Captain James Paton*   (Terra Nova *1913*)
Captain James Paton lived at Rock Ferry, Liverpool. He returned the *Terra Nova* to St John's from Cardiff in August 1913, together with a crew sent from Newfoundland.
(*South Wales Daily News*, August 1913)


*Captain Abram Kean (1855–1945)* (Terra Nova *1906–08, 1920–26, 1932–33*)
Abram Kean was born on 8 July 1855, at Flowers Island, Bonavista Bay, Newfoundland. He was the youngest of nine children in a family of sealers and cod fishers. A legendary figure in the Newfoundland sealing and fishing industry, he began his trade in a sailing schooner and progressed to steam vessels. He married at 17 years of age and had six sons and two daughters. His eldest son was lost from the *Florizel* when aboard as a passenger. In 1885, he entered politics and was elected to the Legislative Council as the member for Bonavista and later held a number of political posts

He had begun his career as a sealing captain in the *Wolf* in 1889, and brought in a record-breaking catch which marked the beginning of a career at the ice which lasted until 1936. He commanded a number of sealers, including the *Terra Nova* and steamers of the Bowring Brothers' Red Cross Line.

Such a long career in a dangerous industry was not without controversy. He is mentioned in the tragedies connected with the SS *Greenland* in 1898

and with the SS *Newfoundland* in 1914, both of which affected his reputation, but men would always sail with him as they always respected his success as a captain. He became known as 'The Admiral of the Fleet'. In 1934 he was awarded the OBE and presented with a model of *Terra Nova* in a glass case at a ceremony held in his honour. A long-term employee of Bowring Brothers Ltd, he died in 1945.
(Kean, Abram, *Old and Young Ahead*, Flanker Press, St John's)

*Captain Edward Bishop (b. 1854)*     (Terra Nova *1909*)
Captain Edward Bishop was born at Swain's Island, Bonavista Bay in 1854, and was captain of sealing steamers for nineteen springs. He married in 1878. A general merchant for forty-five years, he was a Liberal candidate for Bonavista Bay in 1909. He served on church and school boards. He was director and chairman of Wesleyville Mutual Marine Insurance Club.
(*Newfoundland Who's Who*, 1922★)

*Captain William J. Bartlett (1851–1931)*     (Terra Nova *1913–15*)
Born in Brigus, Conception Bay and educated at Harbour Grace Grammar School and Brigus High School, he was the father of Arctic navigator Robert Abram Bartlett. He began fishing at 15, and at 16 accompanied his father to the ice on board SS *Panther*, taking over as captain in 1884 and then for forty-two years he worked in the seal and cod fishing industry at the station established by his father and is credited with making record catches in the industry.
(*Encyclopaedia of Newfoundland and Labrador*, Vol. I★)

*Captain Samuel R. Winsor (b. 1872)*     (Terra Nova *1916*)
Samuel Winsor was a master mariner (inspector of pickled fish and refined cod liver oil). He was born on Swain's Island, Bonavista Bay, educated at the Methodist College and Professor Holloway and married in 1894. He had three children.
(Condon, P.M., *The Seal Fishing of Newfoundland*, 1925★)

*Captain Nicholas J. Kennedy*     (Terra Nova *1917–19*)
Nicholas Kennedy's family resided at Crocker's Cove, Conception Bay. He had two brothers, John and Terence. They were a large family and together

made an annual summer trip to the Labrador, where they fished at Sloop Cove and shared a large home. At the end of the season in 1897, with their fish stocks ready, arrangements made with another ship to convey their return to Newfoundland were misunderstood, leaving Nicholas and his family overstaying in Labrador. They became iced in and made their home with Eskimos at a nearby Bay. But this caused embarrassment to their 17-year-old daughter Julia, for the leading Eskimo who hunted and cared for them wanted her for his wife. The family were eventually relieved by a ship which negotiated the ice.
(*St John's Woman* magazine, September 1963★★)


*Captain Westbury Kean (1886–1974)*     (Terra Nova *1927*)
Born in St John's, Westbury was the son of Abram Kean. He made his first voyage to the Labrador fishery from Brookfield, Bonavista Bay, in 1900 for the family firm. In 1911, he took command of the *Newfoundland*. Though largely exonerated by the inquiry into the *Newfoundland* disaster of 1914, he did not have another command at the ice until *Ranger* from 1921 to 1926. He was master of *Eagle II* for three years from 1930. In 1934, as master of the steamer *Portia* he was accused of trying to smuggle beaver skins out of the country and dismissed. He successfully appealed in 1937 and moved to Halifax, Nova Scotia, as an agent for Imperial Oil. He later returned to Newfoundland and was master of the *Imogene* in 1938 and 1939. He died at New York in 1974.
(*Encyclopaedia of Newfoundland and Labrador*, Vol. III★)


*Captain John Parsons (1868–1949)*     (Terra Nova *alt. 1928–31*)
Born at Bay Roberts, Conception Bay, John Parson was a mariner and politician. He did his navigational training under Captains Isaac Bartlett, Henry Dawe and Charles Dawe and made his first voyage as master in about 1895. He established a wholesale and retail business in the Bay Roberts area, which operated under his name until 1979. From 1906 to 1908 he was captain of the *Newfoundland* at the seal hunt and from 1909 to 1914 he commanded the *Bonaventure* and later the *Erik* and *Diana*. His last two voyages to the seal fishery were in *Terra Nova* in 1929 and 1931. A community and church leader; as a politician he was a strong opponent of confederation with Canada.
(*Encyclopaedia of Newfoundland and Labrador*, Vol. IV★)

*Captain Jacob Kean (1864–1939)*   *(*Terra Nova *alt. 1928–31)*
Born at Flowers Island and educated at Pool's Island and Greenspond, Jacob Kean was married with two children. He was captain of several steamers and made twenty trips to the ice as sealing master involved in a partnership with his uncle, Abram Kean. His first command was *Virginia Lake* in 1907. He left the fishery in 1911 to command the coastal steamer *Home* and remained in the coastal service for the rest of his career, commanding the *Invermore*, *Susu* and *Prospero* until his retirement in 1935. His last trip to the ice was in *Thetis* in 1936.
(*Encyclopaedia of Newfoundland and Labrador*, Vol. III*)


*Captain Stanley G. Barbour (b. 1883)*   *(*Terra Nova *1934–40 & 1941)*
Stanley Barbour was born at Newtown, Bonavista Bay, the son of Captain George Barbour. He took charge of his father's fishing schooners at 19 years of age and worked in the cod fishery for twenty-three years. He was engaged in the seal fishery and in the salmon industry and commanded ships for a number of companies.
(*Newfoundland Who's Who*, 1927*)


*Captain B. Richard Badcock*   *(*Terra Nova *1940)*
He commanded *Ranger* from 1931–33 and earlier the *Viking*,
(Shannon Ryan, *Seals & Sealers*, 1987)


*Captain Llewellyn Lush (1890–1981)*   *(*Terra Nova *1943)*
The last captain of the SS *Terra Nova*, he was born in 1890. During the First World War he served in the Royal Navy as a member of the Royal Navy Volunteer Reserve. He became a marine pilot with the Port of St John's Harbour Authority and retired from this position in 1952. He died on 9 April 1981 and is buried in the Church of England Cemetery at Forest Road, St John's.
(Inf. Tom Goodyear/Mike Tarver)

*For providing this information, I am grateful to the Maritime History Archive, Memorial University of Newfoundland, **and to the Newfoundland Centre for Historical Studies.

# Appendix E

SS *Terra Nova* Crew List, Antarctic Relief Voyage 1903–04

*Voyage to Antarctic to Relieve British National Antarctic Expedition, 1901–04*

| | |
|---|---|
| McKay, Harry | Master |
| Jackson, Alfred P. | Chief Mate |
| Elms, Arthur James | 2nd Mate |
| Day, Roderick Wilson | 3rd Mate |
| Souter, William Clark | Surgeon |
| Sharp, Alexander | Chief Engineer |
| Smith, William | 2nd Engineer |
| McGregor, Colin | 3rd Engineer |
| Aitkin, Alexander | Boatswain |
| Smith, Alexander | Carpenter |
| Smith, Alexander | Carpenter's mate |
| Morrison, Edward | Sailmaker |
| Shearer, Thomas A. | Ass. Steward |
| Grant, John | Cook |
| Clark, William Oliver | Ass. Cook |
| Cairns, James | AB |
| Clark, James | AB |
| Christie, Robert | AB |
| Cosgrove, Thomas | AB |
| Couper, James | AB |
| Dair, John R. | AB |
| Lawrence, George | AB |
| Morrell, Alexander | AB |
| McNeill, Alexander | AB |
| Reilly, James | AB |
| Spaulding, Tasmin | AB |
| Stanistreet, Cyrus | AB |
| Strachan, M. | AB |
| Frederick, David Henderson | AB |
| Frederick, John | Fireman |
| Milne, David T. | Fireman |

(Courtesy of *Antarctica Unveiled* by David Yelverton, 2000)

# APPENDIX F

SS Terra Nova Crew List, Arctic Relief Voyage, 1905

*Voyage to Franz Joseph Land to Relieve Fiala–Ziegler Arctic Expedition, 1903–05*
The following members of the crew for the passage to Franz Joseph Land have been identified from the photograph on page 74. I am grateful to Mr Ivar Stokkeland of the Norsk Polarinstitutt, Tromsø, who made contact with Mr Markus Rotvold of Tromsø, whose father Einar Rotvold was a member of the crew.

(Middle row)
Captain Harry McKay (handing over command)
Mr David Bruce (David Bruce & Co. Shipbrokers)
Captain Johan Kjeld Kjeldsen (taking over command)
Mr William S. Camp with dog (secretary to William Ziegler)

Crew members: Mr Johan Hagerup, Mr Einar Rotvold, Mr Haugen, Mr Morten Olaisen

The remainder have not been identified.

(Courtesy of the Norsk Polarinstitutt, Tromsø)

# Appendix G

SS *Terra Nova*, Shore Party and Crew List
British Antarctic Expedition, 1910–13

| *Officers: Shore Party* | | |
|---|---|---|
| Robert Falcon Scott PP | Captain, CVO RN | (1868–1912) |
| Edward R.G.R. Evans RS | Lieutenant, RN | (1881–1957) |
| Victor L.A. Campbell | Lieutenant, RN ( | Emergency List) (1870–1956) |
| Henry R. Bowers | Lieutenant, Royal Indian Marine | (1883–1912) |
| Lawrence E.G. Oates | Captain, 6th Inniskilling Dragoons | (1880–1912) |
| G. Murray Levick | Surgeon, RN | (1877–1956) |
| Edward L. Atkinson | Surgeon, RN (Parasitologist) | (1881–1929) |

| *Scientific Staff* | | |
|---|---|---|
| Edward A. Wilson | BA, MB (Cantab) Chief Scientific | (1872–1912) |
| George C. Simpson | Staff and Zoologist, DSc Meteorologist | (1878–1965) |
| T. Griffith Taylor | BA, BSc, BE Geologist | (1880–1963) |
| Edward W. Nelson | Biologist | (1883–1923) |
| Frank Debenham | BA, BSc Geologist | (1883–1965) |
| Charles S. Wright | BA, Physicist | (1887–1975) |
| Raymond Priestley | Geologist | (1886–1974) |
| Herbert G. Ponting | FRGS, Camera Artist | (18701935) |
| Cecil H. Meares | In charge of dogs | (1877–1937) |
| Bernard C. Day | Motor Engineer | (1884–1952) |
| Apsley Cherry-Garrard | BA, Assistant Zoologist | (1886–1959) |
| Tryggve Gran | BA, Sub-Lieutenant, Norwegian NR Ski Expert | (1889–1980) |

| Men | | |
|---|---|---|
| William Lashly | Chief Stoker, RN | (1867–1940) |
| William W. Archer | Chief Steward, late RN | (1871–1944) |
| Thomas Clissold | Cook, late RN | (1886–1964) |
| Edgar Evans | Petty Officer, RN | (1876–1912) |
| Robert Forde | Petty Officer, RN | (1887–1959) |
| Thomas Crean | Petty Officer, RN | (1876–1938) |
| Thomas S. Williamson | Petty Officer, RN | (1877–1940) |
| Patrick Keohane | Petty Officer, RN | (1879–1950) |
| George P. Abbott | Petty Officer, RN | (1880–1923) |
| Frank V. Browning | Petty Officer, RN | (1882–1930) |
| Harry Dickason | Able Seaman, RN | (1885–1944) |
| Frederick J. Hooper | Steward, late RN | (1889–1955) |
| Anton Omelchenko | Groom | (1883–1932) |
| Demetri Gerop | Dog Driver | (1889–1932) |
| | | |
| Ship's Party | | |
| Harry L.L. Pennell | Lieutenant, RN | (1882–1916) |
| Henry E. de P. Rennick | Lieutenant, RN | (1881–1914) |
| Wilfred M. Bruce | Lieutenant, RN | (1874–1953) |
| Francis R.H. Drake | Assistant Paymaster, RN (Rtd), Secretary & Meteologist | (1878–1936) |
| Dennis G. Lillie | MA, Biologist | (1884–1963) |
| James R. Dennistoun | In charge of mules (Return voyage) | |
| Alfred B. Cheetham | RNR Boatswain | (1868–1918) |
| William Williams | Chief Engine Room Artificer, RN | (1875–) |
| William A. Horton | Engine Room Artificer, 3rd Class, RN | (–1939) |
| Francis E.C. Davies | Leading Shipwright, RN | (1885–) |
| Frederick Parsons | Petty Officer, RN | (1879–1970) |
| William L. Heald | Petty Officer, late RN | (1876–1939) |
| Arthur S. Bailey | Petty Officer, RN | |
| Albert Balson | Leading Seaman, RN | (1950) |

| | | |
|---|---|---|
| Joseph Leese | Able Seaman, RN | (1884–1948) |
| John H. Mather | Petty Officer, RNVR | (1887–1957) |
| Robert Oliphant | Able Seaman, RN | |
| Thomas F. McLeod | Able Seaman, RN | (1873–1960) |
| Mortimer McCarthy | Able Seaman, RN | (1878–1967) |
| William Knowles | Able Seaman, RN | (1875–) |
| Charles Williams | Able Seaman, RN | (1881–1919) |
| James Skelton | Able Seaman, RNR | (1882–1951) |
| William McDonald | Able Seaman, RN | (1892–1978) |
| James Paton | Able Seaman, RN | (1869–1917) |
| Robert Brissenden | Leading Stoker, RN | (1879–1912) |
| Edward A. McKenzie | Leading Stoker, RN | (1888–1973) |
| William Burton | Leading Stoker, RN | (1889–1988) |
| Bernard J. Stone | Leading Stoker, RN | |
| Angus McDonald | Fireman | (1877–) |
| Thomas McGillon | Fireman | (1885–) |
| Charles Lammas | Fireman | |
| William H. Neale | Steward | |

# APPENDIX H

Summarised Directory of the Arctic Region and Antarctic Continent

I am grateful to Robert K. Headland, historian and archivist for the Scott Polar Research Institute, University of Cambridge, for providing the following data in respect of each polar region.

# THE ARCTIC

The Arctic is essentially an ocean (the smallest of the five oceans), Greenland, the northern extremities of three continents (Asia, Europe, and North America) and several archipelagos. The boundaries are not distinct but are best defined by a combination of the tree-line, the southern limit of continuous permafrost, and the average extent of winter pack ice.

The Arctic Ocean includes the Barents Sea, Kara Sea, Laptev Sea, East Siberian Sea, Chukchi Sea, Beaufort Sea, and Lincoln Sea. It receives a large, although seasonably very variable, fresh-water influx from many of the world's greatest rivers, especially those draining Siberia. Its surface area is $14.5 \times 10^6$ km$^2$ of which a summer minimum of 50% is permanently covered by *pack ice*; this increases to 85% in winter. The mean ice thickness is 3 m; the average duration of a floe is 3 years before it melts or escapes from the Arctic Ocean. At the North Pole ice may drift as much as 20 km daily. The greatest oceanic depth is 5608 m (78°N, 02°W) and the depth at the *North Pole* is 4179 m (the floating ice is about 1·5 m thick). Its submarine topography is complex with several ridges, trenches, abysses, deep and shallow plains. The average depth is 1800 m, thus it is the shallowest of the oceans. Siberia has an extensive continental shelf while that off North America is narrow and drops abruptly. The remote Northern Pole of Inaccessibility (84·05°N, 174·85°W) is 1100 km from the nearest coast (first reached in 1941 by aircraft). The North Magnetic Pole was at 82·2°N, 113·3°W in 2004, off Rolf Ringnes Island; it was first reached in 1831 when farther south.

Greenland, with a surface area of $2.8 \times 10^6$ km$^2$, has the only Arctic *ice sheet* which has an area of $1.8 \times 10^6$ km$^2$, maximum elevation of 3231 m (73°N, 40°W), and a volume of $2.5 \times 10^6$ km$^3$ which is 9% of the ice on Earth. The greatest ice depth measured is 3350 m (72°N, 40°W) and mean depth 1200 m. Bedrock is depressed to a maximum depth of 450 m below sea level in some places. Greenland also has the highest peak in the Arctic; Gunnbjørns Fjæld at 3693 m (68·9°N, 29·9°W, first climbed on 18 August 1935). The many Arctic ice caps are of comparatively minor size; no large ice shelves exist. The most northern land on Earth is Odâq Ø, or a small shingle bank in its vicinity (83·7°N, 30·7°W), off Greenland.

The Arctic has had a peripheral *indigenous population* for many millennia. These include tribes of Eskimo (including Inuit), Lapp, Samoyed, and Chukchi. Abandoned Eskimo settlements have been found beyond 80°N in Greenland. About 80% of the present population of the Arctic immigrated during the 20th century during which urbanization began, nomads settled, and several major population centres were established. *Exploration* of the continental coasts was largely complete almost two centuries ago but many of the islands were charted subsequently. Most of the northern regions of the Canadian Arctic archipelago and Greenland were mapped by 1900. The last major discovery of land on Earth was Severnaya Zemlya in 1913, and the last of the smaller islands was identified in 1947. The *North Pole* was first seen on 12 May 1926, from an airship; first reached by aircraft on 23 April 1948, by a submarine on 3 August 1958, over the pack-ice surface on 6 April 1969, and by icebreaker on 17 August 1977.

*Sealers, whalers, hunters,* and *trappers* have been active in the Arctic for millennia. Extraction of metals, hydrocarbons, and other *minerals* is currently a major economic activity. Many air and some sea routes cross Arctic regions. *Tourism* is minor but increasing. Russia maintains the *Northern Sea Route* (Northeast Passage) for much of the year with icebreakers, some atomic powered. In contrast the *Northwest Passage* has not become a successful commercial waterway.

Many countries conduct research and maintain meteorological observatories and other stations around the Arctic Ocean, on the continents and islands. Drift stations have been deployed on ice floes which circulate around the ocean. A large number of Russian stations have been closed from the early 1990s. Several *military stations* are maintained, both offensive and defensive, although numbers have also decreased greatly after 1990. *Novaya Zemlya* has been the principal Arctic nuclear bomb testing site for the Soviet Union, with 132 detonations from 1955 to 1990; its vicinity has also been used for disposal of radioactive waste. The United States has used sites in Alaska for testing nuclear explosions and four atomic bombs, in a crashed aircraft, spread contamination near Thule, Greenland, in January 1968.

The Arctic *climate* is extreme but the low elevation and proximity of the sea ameliorate it compared with that of the Antarctic. Winds may be severe and precipitation, mainly snow, is generally abundant. The record Arctic minimum temperature is –71·1°C, at Oymyakon, Siberia (63·3°N, 143·2°E), in 1964. The *Aurora Borealis* may be prominent on dark nights.

Arctic *flora* and *fauna* are closely related to those of surrounding continents but have adapted to the harsh climate. *Polar Bears* and Ringed Seals have been found north of 88°N. Migratory birds occupy breeding sites in immense numbers during the brief summer. Barely 5000 years ago Mammoths were present.

Most of the Arctic is unrestricted nationally as 'high seas'. The *Spitsbergen Treaty* (1920) allows access to the Norwegian Svalbard archipelago by the 42 signatory countries. Eight countries govern territory north of the Arctic Circle (66·5°N): Canada, Denmark (for Greenland), Finland, Iceland, Norway, Russia, Sweden, and United States. Coordination of *scientific research* between these, and nearly all other countries involved, is largely through the International Arctic Sciences Committee (founded in 1990), with a Secretariat in Oslo.

R. K. Headland  Scott Polar Research Institute  [Revised 1 April 2005]

# THE ANTARCTIC

The Antarctic includes the continent of *Antarctica* (the fifth largest), the surrounding *Southern Ocean*, and the 19 *peri-Antarctic islands*. The limit of the Southern Ocean is the *Antarctic Convergence* or Polar Front, which is subject to seasonal variation and large gyres. Antarctica is the most isolated of the 7 continents, separate by 900 km from South America, the closest one. It is also the only one entirely isolated by abyssal deeps.

Antarctica is 99·77% covered by a permanent *ice sheet* and has a surface area of $13·88 \times 10^6$ km$^2$, including several massive ice-shelves. It includes sub-continents of Greater Antarctica ($10·35 \times 10^6$ km$^2$) and Lesser Antarctica ($3·53 \times 10^6$ km$^2$), separated by the Transantarctic Mountains. The *South Pole* is 2835 m above sea level (its 'pressure altitude' is about 3300 m), about 2757 m of which is ice. The ice surface at the pole flows about 10 m annually. The mean ice thickness is 2034 m, maximum measured depth is 4776 m (69·9°S, 135·2°E). The ice sheet has a maximum elevation of 4231 m (81°S, 75°E) and a volume of about $25·4 \times 10^6$ km$^3$ which is 90% of that on Earth. Much of the bedrock is depressed below sea level by the weight of ice; the greatest depression is 2496 m (81°S, 110°W). Many peri-Antarctic islands have ice caps, these are of insignificant size compared with the ice sheet.

Antarctica's highest peak is Vinson Massif at 4897 m (78·6°S, 85·4°W, first climbed on 17 December 1966). The remotest spot is the Southern Pole of Inaccessibility (83·8°S, 65·8°E), 1330 km inland (first reached in 1958 by surface traverse). In 2004 the *South Magnetic Pole* was at 64·5°S, 138·2°E, some 50 km off the Terre Adélie coast; it was inland in 1909 when first reached. The average surface elevation of Antarctica is 2300 m, the highest of any continent.

The *Southern Ocean* includes the Scotia Sea, Weddell Sea, Amundsen Sea, Bellingshausen Sea, and Ross Sea. As well as the peri-Antarctic islands there are many submarine seamounts, which also cause important upwellings of nutrient-rich waters. Much of the ocean is covered by *pack ice* which has an average winter maximum area of $20 \times 10^6$ km$^2$ (60% of the Southern Ocean). At the height of summer this decreases to: $2·4 \times 10^6$ km$^2$ (12% of the ocean). The mean thickness of the ice is 1·2 m; the average duration of a floe is only 1 year as the majority drift north during summer. Massive tabular icebergs, calved from the ice shelves, are a distinctive feature of the ocean. Its greatest measured depth is 8325 m (56°S, 26°W).

The *first lands* seen in the Antarctic were several of the peri-Antarctic islands. Mainland Antarctica was first sighted in 27 January 1820 and the first landing was probably in late 1820. It was not until 1899 that a winter was spent on the continent and continuous presence began in 1944. The *South Pole* was reached on 14 December 1911. *Sealers*, mainly during the 19th century, and *whalers* during the 20th century were major exploiters of Antarctic resources. The latter established land stations on several of the peri-Antarctic islands. Currently *fishing* and *tourism* are the only commercial operations; research is the principal activity. Unlike in the Arctic neither mineral extraction nor commercial transport routes exist in the Antarctic. Many economic minerals undoubtedly occur but none have yet been exploited.

During the 2004 austral winter 43 *stations* were open in Antarctic regions (25 on Antarctica, all but 2 on coasts) recording meteorological data and involved in other scientific research. These were operated by 18 countries. The winter population of the Antarctic is about 1100, at least twice as many are present during the brief summer. Most stations are relieved by icebreakers and other ships. Only five intercontinental landing strips are maintained.

The Antarctic *climate* is the most severe on Earth. Winds often become blizzards and a minimum temperature of −89·2°C has been recorded (Earth's lowest) at Vostok station (78·4°S, 106·9°E) in 1983. The continent is essentially a frigid desert because there is very little precipitation from the cold dry atmosphere - and virtually all this is frozen (snow). In many areas sublimation may exceed melting in ablating glaciers, and humidifying the air. The *Aurora Australis* may be prominent on dark nights.

Terrestrial *flora* and *fauna* are highly endemic. They are characterized by few species which may occur in very isolated concentrations. A few lichens and algae survive on the most remote nunataks. Marine organisms are, on the contrary, abundant in local situations and include many species of whales, seals, and commercial fish. The peri-Antarctic islands are particularly important breeding sites for *penguins* and other sea-birds, and seals. No indigenous humans have existed; although a small number of children have been born on continental Antarctica (the first was in 1978), and on several peri-Antarctic islands.

The part of the Antarctic south of 60°S is subject to the *Antarctic Treaty* (made in 1959) which, currently, has 45 signatory countries (covering over 80% of the Earth's population). It has a Secretariat in Buenos Aires, established in 2004. The treaty puts the 7 sovereign claims (Argentina, Australia, Britain, Chile, France, New Zealand, and Norway, some of which overlap) in abeyance and its subsequent instruments regulate most human activities. Most current *military activities* in the Treaty region are to provide transport for, and supplies to, the stations, and for hydrographic survey. Neither nuclear explosions nor disposal of radioactive waste have been reported within the Treaty region (although nuclear explosions and military operations have occurred in other parts of the Antarctic). *Scientific research* is coordinated internationally by the Scientific Committee on Antarctic Research (founded in 1958), with a Secretariat in Cambridge, and its specialised groups.

R. K. Headland     Scott Polar Research Institute     [Revised 1 April 2005]

Scott Polar Research Institute

## Antarctic Winter Stations 2004

R. K. Headland.  1 July 2004

During the 2004 austral winter 43 stations were open in Antarctic regions recording meteorological data and involved in other scientific research. These were operated by 18 countries. In the list the symbol • indicates the station is on one of the 19 peri-Antarctic islands (18 stations), and * that it is in the region covered by the Antarctic Treaty (35 stations - 10 also marked •). The number in brackets indicates the position clockwise from the prime meridian. Some stations are in ice shelves which move, thus their geographical coordinates are somewhat variable; these are marked with the sign ≈.

### National Order

*Argentina*
 'Esperanza', Hope Bay *; 63·36°S, 56·99°W   *[34]*
 'Belgrano II', Coats Land *; 77·87°S, 34·63°W   *[39]*
 'San Martín', Barry Island *; 68·13°S, 67·10°W   *[21]*
 'Orcadas', Laurie Island *•; 60·74°S, 44·73°W   *[36]*
 'Jubany', King George Island *•; 62·23°S, 58·67°W   *[30]*
 'Marambio', Seymour Island *; 64·24°S, 56·66°W   *[35]*

*Australia*
 Macquarie Island •; 54·50°S, 158·95°E   *[17]*
 'Casey', Vincennes Bay *; 66·28°S, 110·52°E   *[15]*
 'Davis', Ingrid Christensen Coast *; 68·57°S, 77·97°E   *[12]*
 'Mawson', Mac. Robertson Land *; 67·60°S, 62·87°E   *[7]*

*Brasil*
 'Comandante Ferraz', King George Island *•; 62·08°S, 58·39°W   *[32]*

*Britain*
 Bird Island, South Georgia •; 54·00°S, 38·05°W   *[37]*
 King Edward Point, South Georgia •; 54·28°S, 36·50°W   *[38]*
 'Halley', Brunt Ice Shelf *; ≈75·58°S, ≈26·54°W   *[40]*
 'Rothera', Adelaide Island *; 67·57°S, 68·12°W   *[20]*

*Chile*
 'Capitán Arturo Prat', Greenwich Island *•; 62·50°S, 59·68°W   *[24]*
 'General Bernardo O'Higgins Riquelme', Cape Legoupil *; 63·32°S, 57·90°W   *[33]*
 'Presidente Eduardo Frei Montalva', King George Island *•; 62·21°S, 58·97°W   *[26]*

*China (People's Republic)*
 'Chang Cheng' ['Great Wall'], King George Island *•; 62·21°S, 58·96°W   *[25]*
 'Zhongshan' ['Sun Yat-sen'], Larsemann Hills *; 69·37°S, 76·37°E   *[10]*

*France*
 Port-aux-Français, Iles Kerguelen •; 49·35°S, 70·20°E   *[8]*
 'Alfred-Faure', Iles Crozet •; 46·43°S, 51·87°E   *[6]*
 'Dumont d'Urville', Terre Adélie *; 66·67°S, 140·01°E   *[16]*
 'Martin-de-Viviès', Ile Amsterdam •; 37·83°S, 77·59°E   *[11]*

*Germany*
 'Neumayer', Ekstrømisen *; ≈70·65°S, ≈08·25°W   *[42]*

*India*
 'Maitri', Shirmacheroasen *; 70·77°S, 11·73°E   *[2]*

*Japan*
 'Syowa', Ongul *; 69·00°S, 39·58°E   *[5]*

*Korea (Seoul)*
 'King Sejong', King George Island *•; 62·22°S, 58·79°W   *[29]*

*New Zealand*
 'Scott Base', Ross Island *; 77·85°S, 166·76°E   *[19]*

*Poland*
 'Henryk Arctowski', King George Island *•; 62·17°S, 58·47°W   *[31]*

*Russia*
 'Bellingsgausen', King George Island *•; 62·20°S, 58·97°W   *[27]*
 'Mirnyy', Queen Mary Land *; 66·55°S, 93·01°E   *[13]*
 'Progress', Larsemann Hills *; 69·38°S, 76·39°E   *[9]*
 'Novolazarevskaya', Shirmacheroasen *; 70·77°S, 11·87°E   *[3]*
 'Vostok', South Geomagnetic Pole *; 78·47°S, 106·80°E   *[14]*

*South Africa*
  Gough Island •; 40·36°S, 09·87°W  *[41]*
  Marion Island •; 46·87°S, 37·86°E  *[4]*
  'SANAE IV', Vesleskarvet *; 71·68°S, 02·83°W  *[43]*

*Ukraine*
  'Akademician Vernadskiy', Argentine Islands *; 65·25°S, 64·26°W  *[22]*

*United States*
  'Amundsen-Scott', South Pole *; 90°S  *[1]*
  'McMurdo', Ross Island *; 77·85°S, 166·67°E  *[18]*
  'Palmer', Anvers Island *; 64·77°S, 64·05°W  *[23]*

*Uruguay*
  'Artigas', King George Island *•; 62·18°S, 58·90°W  *[28]*

**Geographical Order**
[From the South Pole and eastwards from the prime meridian.]

*1* 'Amundsen-Scott' (United States), South Pole *; 90°S
*2* 'Maitri' (India), Shirmacheroasen *; 70·77°S, 11·73°E
*3* 'Novolazarevskaya' (Russia), Shirmacheroasen *; 70·77°S, 11·87°E
*4* Marion Island (South Africa) •; 46·87°S, 37·86°E
*5* 'Syowa' (Japan), Ongul *; 69·00°S, 39·58°E
*6* 'Alfred-Faure' (France), Iles Crozet •; 46·43°S, 51·87°E
*7* 'Mawson' (Australia), Mac. Robertson Land *; 67·60°S, 62·87°E
*8* Port-aux-Français (France), Iles Kerguelen •; 49·35°S, 70·20°E
*9* 'Progress' (Russia), Larsemann Hills *; 69·40°S, 76·37°E
*10* 'Zhongshan' ['Sun Yat-sen'] (China), Larsemann Hills *; 69·37°S, 76·37°E
*11* 'Martin-de-Viviès' (France), Ile Amsterdam •; 37·83°S, 77·59°E
*12* 'Davis' (Australia), Ingrid Christensen Coast *; 68·57°S, 77·97°E
*13* 'Mirnyy' (Russia), Queen Mary Land *; 66·55°S, 93·01°E
*14* 'Vostok' (Russia), South Geomagnetic Pole *; 78·47°S, 106·80°E
*15* 'Casey' (Australia), Vincennes Bay *; 66·28°S, 110·51°E
*16* 'Dumont d'Urville' (France), Terre Adélie *; 66·67°S, 140·02°E
*17* Macquarie Island (Australia) •; 54·50°S, 158·95°E
*18* 'McMurdo' (United States), Ross Island *; 77·85°S, 166·67°E
*19* 'Scott Base' (New Zealand), Ross Island *; 77·85°S, 166·76°E
*20* 'Rothera' (Britain), Adelaide Island *; 67·57°S, 68·12°W
*21* 'San Martín' (Argentina), Barry Island *; 68·13°S, 67·10°W
*22* 'Akademician Vernadskiy' (Ukraine), Argentine Islands *; 65·25°S, 64·26°W
*23* 'Palmer' (United States), Anvers Island *; 64·77°S, 64·05°W
*24* 'Capitán Arturo Prat' (Chile), Greenwich Island *•; 62·50°S, 59·68°W
*25* 'Chang Cheng' ['Great Wall'] (China), King George Island *•; 62·21°S, 58·96°W
*26* 'Presidente Eduardo Frei Montalva' (Chile), King George Island *•; 62·21°S, 58·97°W
*27* 'Bellingsgausen' (Russia), King George Island *•; 62·20°S, 58·97°W
*28* 'Artigas' (Uruguay), King George Island *•; 62·18°S, 58·90°W
*29* 'King Sejong' (Korea [Seoul]), King George Island *•; 62·22°S, 58·79°W
*30* 'Jubany' (Argentina), King George Island *•; 62·23°S, 58·67°W
*31* 'Henryk Arctowski' (Poland), King George Island *•; 62·17°S, 58·47°W
*32* 'Comandante Ferraz' (Brasil), King George Island *•; 62·08°S, 58·39°W
*33* 'General Bernardo O'Higgins Riquelme' (Chile), Cape Legoupil *; 63·32°S, 57·90°W
*34* 'Esperanza' (Argentina), Hope Bay *; 63·36°S, 56·99°W
*35* 'Marambio' (Argentina), Seymour Island *; 64·24°S, 56·66°W
*36* 'Orcadas' (Argentina), Laurie Island *•; 60·74°S, 44·73°W
*37* Bird Island (Britain), South Georgia •; 54·00°S, 38·05°W
*38* King Edward Point (Britain), South Georgia •; 54·28°S, 36·50°W
*39* 'Belgrano II' (Argentina), Coats Land *; 77·87°S, 34·63°W
*40* 'Halley' (Britain), Brunt Ice Shelf *; ≈75·58°S, ≈26·54°W
*41* Gough Island (South Africa) •; 40·36°S, 09·87°W
*42* 'Neumayer' (Germany), Ekstrømisen *; ≈70·65°S, ≈08·25°W
*43* 'SANAE IV' (South Africa), Vesleskarvet *; 71·68°S, 02·83°W

# Appendix I

## Alexander Stephen & Sons Ltd, 1750–1970: The Family Line of Shipbuilders and a Brief History of the Company

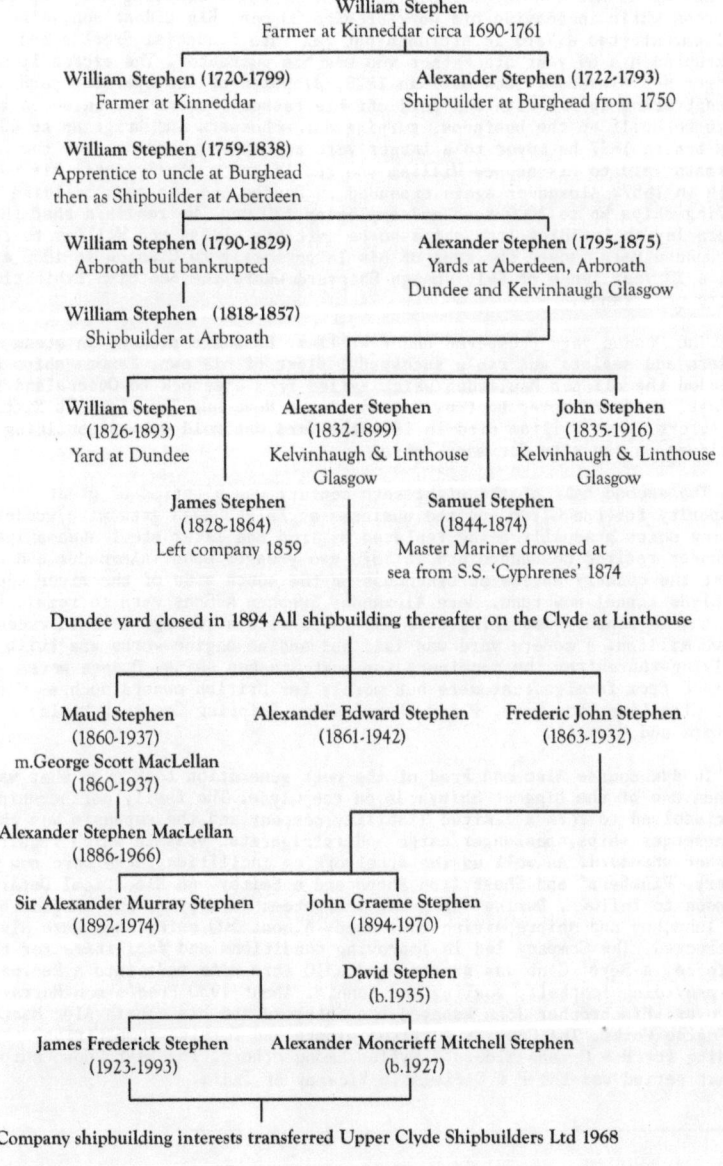

ALEXANDER STEPHEN & SONS LTD - A BRIEF HISTORY

Alexander Stephen, born of farming stock in 1722, started a small shipbuilding and repairing business at Burghead near Elgin about 1750. The business was carried on successfully by his son and grandson but very little is known about it. Alexander's nephew William was apprenticed to him and set up on his own in Aberdeen in 1793. He built coasting ships up to 200 tons while increasing his workforce to 30 men. His eldest son, also William, started a Yard in Arbroath but ran into financial trouble and bankrupted his 69 year old Father who was his guarantor. The extremely able younger son Alexander took over in 1828, disposed of the Aberdeen yard concentrating on Arbroath and paid off his Father's creditors in seven years. There he built up the business, turning out schooners and brigs up to 400 tons but in 1837 he moved to a larger yard at Dundee, handing over the Arbroath yard to his nephew William who ran it successfully until his untimely death in 1857. Alexander again expanded in Dundee and was soon building sailing ships up to 1000 tons and employing 130 men. He realised that the future lay in building iron ships so he left his eldest son William to run the Dundee yard, moved the rest of his large family to Glasgow in 1850 and took a 20 year lease of Kelvinhaugh Shipyard where the Scottish Exhibition Centre now stands.

The Dundee yard prospered under William. He was a pioneer in steam whalers and sealers and ran a successful fleet of his own. Famous ships built included the clipper Maulesden which sailed from Greenock to Queensland in 70 days, a record never beaten, Scott's Terra Nova and Shackleton's Nimrod and Aurora. When William died in 1894 the Yard was sold but shipbuilding continued on the site for many years.

The second half of the eighteenth century was a period of great prosperity for the Clyde and the business at Kelvinhaugh grew with wooden sailing ships gradually being replaced by iron and later steel steamships. Alexander retired to Dundee and in 1870 two younger sons, Alexander and John, bought the country estate at Linthouse on the south side of the river where the Clyde tunnel now runs. Here Alexander Stephen & Sons were to remain for more than a century launching over 550 ships with a tonnage well in excess of two million. A modern yard was laid out and an engine works was built supplying thereafter the machinery for most Stephen ships. Orders were obtained from foreign customers but mostly for British owners such as City Line, Clan Line, Ben Line, P & O, New Zealand Shipping Company, Maclay McIntyre and Burrell.

In due course Alec and Fred of the next generation took over what was by then one of the biggest Shipyards on the Clyde. The family partnership was dissolved to form a limited liability company and the emphasis was changed to passenger ships, passenger/cargo and refrigerated vessels which required a higher standard. As well as the steelworking facilities there were now Joinery, Plumbers' and Sheet Iron Shops and a Smithy. An Electrical Department was soon to follow. During World War 1 eighteen destroyers and torpedo boats were launched and shiprepairing increased. Almost 500 aeroplanes were also constructed. The Company led in improving conditions and facilities for the workforce. A Boys' Club was started and a 10 acre site made into a Recreation Club providing football, bowling and tennis. About 1930 Fred's son Murray took over. His brother John managed the Shipyard and his cousin Alec MacLellan the Engine Works. The Company survived slumps and foreign competition, building for P & O and Elders & Fyffes among others. The best known ship of that period was the P & O flagship Viceroy of India.

# Appendix I

World War 2 saw the Yard building an aircraft carrier, cruisers, destroyers, frigates, and corvettes. Many achieved great things but the best known was probably HMS Amethyst whose exploits were recorded in the film 'The Yangtse Incident'. Building of merchant ships continued unabated and shiprepairing again became vitally important.

When peace came the reconversion of troop carriers to passenger ships occupied a large number of men, and at one time over 5,000 were employed by the Company. Production soon returned to normal although passenger ships were fewer due to competition by aeroplanes. Cargo liners, refrigerated ships and the occasional warship and tanker became the core of the business. In 1947 Alexander Stephen & Sons Ltd became a public Company with a stock exchange quotation, although the family kept forty percent of the shares. In 1950 the Bicentenary was celebrated by publishing the excellent history 'Stephen of Linthouse' and the introduction of a non-contributory pension scheme for long service hourly paid employees.

Immediately after the war the Directors set about modernising the facilities, changing from rivetting to welding and steel prefabrication. Alexander Stephen & Sons soon led the country. However the entrenched position of the Trade Unions and the power that they wielded caused resistance at all stages so that the benefits could never be fully achieved and productivity was virtually static. By 1960 the Japanese were quoting prices about thirty percent below British prices and were improving their efficiency all the time. The writing was on the wall for British and indeed European shipbuilding. In spite of the looming problems Stephens continually strove to modernise production, introducing computers for accounting and planning.

Sir Murray and John Stephen retired about 1960 and Murray's son Jim took over with his younger brother Sandy sharing some of the responsibility. Adding dredgers to the shipbuilding range helped to fill the Yard but in 1968 the Shipyard was merged reluctantly with four other Yards to form Upper Clyde Shipbuilders. The merger turned out to be a disaster and the group was bankrupt within four years. Stephens were left with their Engine Works and Shiprepairing Department together with a few other small and generally not very profitable activities. Thus Alexander Stephen & Sons ceased shipbuilding although a tin dredger for Bolivia and a bucket dredger for India, both erected on site, were delivered in the next three years. The Engine Works built ships' stabilisers, parts for power stations and much other subcontract work and made dramatic improvements in productivity. However, the lack of steady work and the nationwide falling off of heavy engineering meant the Works had to close in 1976. The shiprepairing side also made considerable strides and overcame the lack of a supporting shipyard but the continuing decline of the Clyde as a port and the decrease in traffic to the west coast of Britain meant a closure the same year. The property was sold by 1982 and Alexander Stephen & Sons Ltd went into voluntary liquidation.

Stephen ships still sail the seas although they are becoming fewer in number and a few are preserved in Maritime Museums. Several took part in the Falklands campaign and the Fleet Auxiliary Sir Galahad was damaged beyond repair in Bluff Cove. The Engine Shop, built in 1870, was declared an historic monument, was dismantled and re-erected in 1992 at Irvine, Ayrshire to form the centre of the Scottish Maritime Museum. Apart from these Alexander Stephen and Sons is now history, but a history of which very many can still be proud.

# Appendix J

Bowring Brothers Ltd, Profit & Loss Accounts Balance Sheet, 1943

## C. T. BOWRING & CO. LIMITED
### LONDON & LIVERPOOL

MANAGERS OF
THE BOWRING STEAMSHIP C° L?°
THE BEARCREEK OIL & SHIPPING C° L?°

CORRESPONDENTS
BOWRING BROS. LTD
ST. JOHNS, NEWFOUNDLAND
BOWRING & CO
NEW YORK

TELEPHONE: ROYAL 7200 (40 LINES)
TELEGRAPHIC ADDRESS.
BOWBOTES, LONDON, TELEX 21256/7

52, LEADENHALL STREET

LONDON, E.C.3

PLEASE ADDRESS REPLY

SHIP MANAGEMENT DEPT.

---

BOWRING BROTHERS, LIMITED

BALANCE SHEET

PROFIT AND LOSS ACCOUNTS

AND

SUNDRY SCHEDULES

FOR THE YEAR ENDED

31st DECEMBER, 1943

ST. JOHN'S,
NEWFOUNDLAND.

## BOWRING BROTHERS LIMITED.

Exhibit "A".

### STEAMERS PROFIT AND LOSS ACCOUNT FOR THE YEAR ENDED 31st. DECEMBER, 1943.

PROFIT ON:-

| | | | |
|---|---:|---:|---:|
| S. S. "Terra Nova", charter to Newfoundland Base Contractors | | | 4,899.76 |
| S. S. "Eagle"- | | | |
|    Charter to Newfoundland Base Contractors | | 16,241.53 | |
|    Do. McNamara Construction Company | | 1,794.51 | |
|    Trip to North Sydney for Bunker Coal | | 5.95 | |
| | | 18,041.99 | |
|    LESS Expenses- Salvage Trips (No assistance rendered) | | 1,055.67 | 16,986.32 |
| | | | 21,886.08 |
| Water Boat "Shamrock" | | | 8,733.22 |
| | | | 30,619.30 |
| LESS | | | |
| LOSS ON:- | | | |
| Lighters and Motor Boats- Excess of Expenses over Revenue | | | 466.08 |
| **PROFIT CARRIED TO GENERAL PROFIT AND LOSS ACCOUNT** | | | **$30,153.22** |

NOTE:-
    Depreciation appears in Exhibit "E".

## BOWRING BROTHERS LIMITED.

Schedule No. 7.

### SEALING PROPERTY 31st. DECEMBER, 1943.

| | Balance. | Sales (Old). | Depreciation. | Balance 31st.Decr. 1943. |
|---|---|---|---|---|
| S. S. "Eagle" | 11,500.00 | - - | 3,500.00 | 8,000.00 |
| Sealoil Manufacturing Plant - Cost | 47,451.33 | 2,500.00 | 32,000.00 | 12,951.33 |
| Sealskinning Machinery- Cost | 19,752.15 | - - | 17,952.15 | 1,800.00 |
| Seal Factory | 10,000.00 | - - | 2,000.00 | 8,000.00 |
| Seal Factory Boilers- Cost | 5,954.16 | - - | 4,166.12 | 1,788.04 |
| Sunning Loft- Cost | 2,627.33 | - - | 1,827.33 | 800.00 |
| Circular Oil Storage Tanks | 3,500.00 | - - | 350.00 | 3,150.00 |
| Steam Boiler for Hoisting Seals- Cost | 692.98 | - - | 277.20 | 415.78 |
| Electric Oil Pump- Cost | 598.41 | - - | 538.56 | 59.85 |
| Sealskins Dressing Machinery- Cost | 2,459.24 | - - | 1,866.47 | 592.77 |
| | $104,535.60 | $2,500.00 | $64,477.83 | $37,557.77 |

S. S. "Terra Nova"  $12,500.00 Total Loss September 12th. 1943- See Schedule No. 19, Steamers' Replacement Reserve.

## BOWRING BROTHERS LIMITED.

Schedule No. 19.

### RESERVES      31st. DECEMBER, 1943.

**GENERAL:-**
| | | |
|---|---:|---:|
| As per last Account | 350,000.00 | |
| Add Transfer from Profit and Loss Appropriation Account | 150,000.00 | 500,000.00 |

**FOR CONTINGENCIES:-**
| | | |
|---|---:|---:|
| As per last Account | 350,000.00 | |
| Add Transfer from Profit and Loss Appropriation Account | 50,000.00 | 400,000.00 |

**SUNDRY:-**

| | | | |
|---|---:|---:|---:|
| Private Insurance, as per last Account | | 3,500.00 | |
| Plate Glass Insurance, as per last Account | 2,932.48 | | |
| Add Premiums for 1943 | 217.49 | | |
| | 3,149.97 | | |
| Less Claims settled | 25.50 | 3,124.47 | |
| Workmen's Compensation, as per last Account | 2,287.38 | | |
| Add Premiums for 1943 | 1,369.09 | | |
| | 3,656.47 | | |
| Less Claims paid 1943 | 577.47 | 3,079.00 | 9,703.47 |

**STEAMER REPLACEMENT:-**

| | | | |
|---|---:|---:|---:|
| As per last Account | | 282,682.43 | |
| ADD- | | | |
| S.S. "Terra Nova"- Claim on Underwriters for total loss of Vessel off Hollander Island, East Coast of Greenland, 12th. September, 1943. | | | |
| Insured Value | 60,000.00 | | |
| Less 1% Collecting Commissions | 600.00 | | |
| | 59,400.00 | | |
| Less- | | | |
| Balance as per last Account, Hull and Machinery-Valued    12500.00 | | | |
| Less- | | | |
| Depreciation 1943    1750.00 | 10,750.00 | | |
| | 48,650.00 | | |
| Less Cost of 2 Lifeboats placed on board as additional Equipment while on charter (lost with ship) | 3,345.00 | 45,305.00 | 327,987.43 |

$1,237,690.90

# Appendix K

## Some Sealing Phrases and Expressions

| | |
|---|---|
| Chew de fat | To Talk |
| Crop | Sealers outfit |
| Dog | To follow |
| Fat | Seals; sculps |
| Greasy jackets | Sealers |
| Harps | Seals of the species *Phoca groenlandica* |
| Hoods | Seals of the species *Cystophora cristota* (larger and more aggressive) |
| High-liner | Leading ship or man in number of seals or fish taken |
| Live swiles | Freshly killed seals |
| Lobscouse | Kind of stew |
| Pan | A cake of ice; a heap or pile of skins; to gather skins into pans |
| Pierhead jump | A quick start |
| Raftering (Nfld) | Large thick sheets of ice doubling up, driven by winds and currents (rafting) |
| Skulp | To skin a seal; the skin and fat of a seal |
| Scun | To direct a ship's course from aloft |
| Scunner | A man who scuns |
| Slob | Thin, loose ice |
| Stick | Speed |
| Swile | To hunt seals |
| Swill | Swell in the ice |
| Whitecoat | A young harp seal |

# Appendix L

Miscellaneous List of Whalers and Sealers Launched and Their Fates

The list of ships below, owned by many companies, including C.T. Bowring & Co., is compiled from a number of sources.

| Wooden ships | Builder | Launched | Lost |
|---|---|---|---|
| *Algerine* | not known | 1893 | 1912 |
| *Arctic* | Alexander Stephen & Sons | 1867 | 1854 |
| *Arctic II* | Alexander Stephen & Sons | 1875 | 1887 |
| *Ariel* | not known | 1873 | Sold 1890 |
| *Aurora* | Alexander Stephen & Sons | 1876 | 1917/18 |
| *Bear* (formerly *Bear of Oakland*) | Alexander Stephen & Sons | 1874 | 1963 |
| *Bloodhound II* | Alexander Stephen & Sons | 1872 | 1917 |
| *Commodore* | Alexander Stephen & Sons | 1871 | 1883 |
| *Diana* | Alexander Stephen & Sons | 1870 | 1922 |
| *Discovery* | Dundee Shipbuilders Ltd | 1901 | Still afloat at Dundee |
| *Endurance* (*Polaris*) | Norwegian | 1913 | 1915 |
| *Erik* | Alexander Stephen & Sons | 1865 | 1918 |
| *Esquimaux* (*America*) | Alexander Stephen & Sons | 1865 | 1903 |
| *Eagle I* | Not known | 1818 | – |
| *Eagle II* | Alexander Stephen & Sons | 1871 | 1893 |

| Wooden ships | Builder | Launched | Lost |
|---|---|---|---|
| Eagle III (Sophie) | Norwegian | 1902 | 1950 |
| Fram | Colin Archer, Larvick, Norway | 1892 | Museum, Oslo |
| Grand Lake | Not known | 1892 | 1908 |
| Greenland | Aberdeen, Scotland | 1872 | 1907 |
| Hawk (ex. HMS Plover) | Not known | Acquired 1865 | 1876 |
| Hope | A. Hall & Co., Aberdeen | 1873 | 1901 |
| Iceland | Alexander Stephen & Son | 1872 | 1910 |
| Jean of Bo'ness | Not known | 1803 | – |
| Kite | Not known | 1873 | 1913 sold |
| Labrador | Not known | – | 1913 |
| Leopard | Ayr, Scotland | 1873 | 1907 |
| Lion | Not known | 1867 | 1882 |
| Mastiff | Alexander Stephen & Sons | 1867 | 1898 |
| Merlin | Not known | 1850 | 1882 |
| Micmac | Not known | – | 1878 |
| Monticello | Not known | – | 1872 |
| Magnific | Not known | 1899 | 1903 sold |
| Neptune | Alexander Stephen & Sons | 1872 | 1943 |
| Narwal | Alexander Stephen & Sons | 1859 | 1884 |
| Newfoundland | Not known | 1872 | 1916 |
| Nimrod | Alexander Stephen & Sons | 1866 | 1919 |
| Osprey | Not known | – | 1874 |
| Panther | Not known | 1861 | 1908 |
| Polynia | Not known | 1861 | 1891 |

| Wooden ships | Builder | Launched | Lost |
|---|---|---|---|
| *Proteus* | Alexander Stephen & Sons | 1873 | 1883 |
| *Ranger* | Alexander Stephen & Sons | 1871 | 1942 |
| *Resolute* | Alexander Stephen & Sons | 1880 | 1886 |
| *Retriever* | Alexander Stephen & Sons | 1865 | 1872 |
| *Scotia (Heckla)* | Norwegian | 1872 | 1916 |
| *Southern Cross (Pollux)* | Norwegian\Colin Archer | 1886 | 1914 |
| *Terra Nova* | Alexander Stephen & Sons | 1884 | 1943 |
| *Thetis* | Alexander Stephen & Sons | 1881 | 1936 |
| *Tiger* | Not known | – | 1884 |
| *Tigress* | Not known | – | 1875 |
| *Vanguard* | Not known | 1872 | 1909 |
| *Viking* | Norwegian | 1881 | 1931 |
| *Virginia Lake* | Not known | – | 1909 |
| *Walrus* | Not known | 1870 | 1908 |
| *Wolf I* | Alexander Stephen & Sons | 1863 | 1871 |
| *Wolf II* | Alexander Stephen & Sons | 1871 | – |
| *Xanthus* | Not known | – | 1880 |
| *Young Harp* | Not known | 1926 | 1956 |

| Steel ships | Builder | Launched | Lost |
|---|---|---|---|
| *Adventure* | Dundee Shipbuilders Co. | 1905 | 1916 Sold to Russia |
| *Bellaventure* | D & W Henderson, Glasgow | 1909 | 1916 Sold to Russia |
| *Bonaventure* | Napier Miller & Co., Glasgow | 1909 | 1916 Sold to Russia |
| *Beothic* | American Steamboat Co. | 1918 | 1940 |
| *Caribou* | Rotterdam, Holland | 1925 | 1942 Torpedoed |
| *Florizel* | Charles Connell & Co., Glasgow | 1909 | 1918 |
| *Sagona* | Dundee Shipbuilding Co. | 1912 | 1940 Sold |
| *Stephano* | Port Glasgow | 1911 | 1916 Torpedoed |
| *Imogene* | Not known | 1929 | 1940 |
| *Kyle* | Not known | 1913 | 1965 |
| *Nascopie* | Not known | 1912 | 1947 |

# REFERENCES

(NB: publishing details, where available, can be found in the bibliography.)

## Chapter 1 – Into the Evening of a Passing Age

Watson, Norman, *The Dundee Whalers*.
Linklater, Eric, *Voyages of the Challenger*.
Rice, Tony, *British Oceanographic Vessels 1800–1950*.
Savours, Ann, *Voyages of the Discovery*.
Carvel, J.L., *Stephen of Linthouse*.
Stephen, A.M.M., Stephen family of shipbuilders from the family tree, University of Glasgow archives.
National Maritime Museum, Greenwich.
*The Dundee Advertiser*, 30 December 1884.
Memo and letter from Bowring Brothers Ltd to H.S. Richards, Swansea Museum, 16 July 1962.
Company sealing records, Bowring Brothers Ltd, University of Glasgow archives.
Keir, David, *The Bowring Story*.
Middlemiss, Norman L., *British Shipbuilding Yards*.

## Chapter 2 – Across the Atlantic and New Owners

*The Ambitions and Achievements of Benjamin Bowring and his Family*
Keir, David, *The Bowring Story*.
Wardle, Arthur C., *Benjamin Bowring and his Descendants*.
Ryan, Shannon & Martha Drake, *Seals and Sealers*.
*International Directory of Company Histories 1991*, Vol. 2, p.82.
Memorandum from Bowring Brothers, Ltd, University of Glasgow archives.
Fleming, Fergus, *Ninety Degrees North*.
Sobel, Dava, *Longitude*.
Reader's Digest, *Antarctica: Great Stories from the Frozen Continent*.

## Chapter 3 – From the North Atlantic to the Antarctic

Scott, Captain R.F., *The Voyage of the* Discovery, Vol. 1 & 2
Yelverton, David, *Antarctica Unveiled*.
*The Dundee Evening Telegraph* (1951).
*The Courier* (Dundee, 1919).


## Chapter 4 – From the Antarctic to the Arctic

Holland, Clive, *Arctic Chronology*.
Fiala, Anthony, *Fighting the Polar Ice*.
Fleming, Fergus, *Ninety Degrees North*.
Extracts from the logbook of Ejnar Mikkelsen (Baldwin–Ziegler Arctic Expedition 1901–02).
*Dukes County Intelligencer*, 1998, Vol. 40, Nos 1 & 2, and 1999, Vol. 40, Nos 3 & 4.
Reader's Digest, *Antarctica: Great Stories from the Frozen Continent*.
Extract from the diary of F.L. Andrassen of the *Belgica*, 1905, per Kjell Kjar extracts from the logbook of Ejnar Mikkelsen and quoting the diary of Captain Kjeld Kjeldsen.
Sealing records of Bowring Brothers Ltd, University of Glasgow archives.


## Chapter 5 – Her Name will be Remembered Forever

Yelverton, David, *Antarctica Unveiled*.
Huntford, Roland, *Scott and Amundsen*.
Evans, Admiral E.R.G.R., *South with Scott*.
Johnson, Dr Anthony M., *Scott of the Antarctic and Cardiff*.
Chappell, Edgar L., *History of the Port of Cardiff*.
Jenkins, J. Geraint, & David Jenkins, *Cardiff Shipowners*.
Gwynn, Stephen, *Captain Scott*.
Ballinger, John, *A Guide to Cardiff, City and Port*.
Wilson, D.M., & D.B. Elder, *Cheltenham in Antarctica: The Life of Edward Wilson*.
*The Western Mail*, 1910.
Limb, Sue, & Patrick Cordingly, *Captain Oates: Soldier and Explorer*.

Wilson, Edward, *Diary of the Terra Nova Expedition to the Antarctic 1910–1912* (Littlehampton Book Services Ltd, 1972).
Cherry-Garrard, Apsley, *The Worst Journey in the World*.
Hattersley-Smith, Geoffrey & Ellen Johanne McGhie (née Gran), *The Norwegian with Scott: Tryggve Gran's Antarctic Diary 1910–1913*.
Bull, Colin, & Pat F. Wright (eds), *'Silas': The Antarctic Diaries and Memoir of Charles S. Wright* (quoting the diary of G.C. Simpson).
Pound, Reginald, *Scott of the Antarctic*.
Ludlam, Harry, *Captain Scott: the Full Story*.
Goodenough, Clive, *History of the Port of Lyttelton*.
Church, Ian, *Last Port to Antarctica: Dunedin and Port Chalmers: 100 years of Polar Service*.
Ponting, Herbert, *The Great White South*.

## Chapter 6 – Into the Southern Ocean

Church, Ian, *Last Port to Antarctica: Dunedin and Port Chalmers: 100 years of Polar Service*.
Evans, Admiral E.R.G.R., *South with Scott*.
Huxley, Leonard, *Scott's Last Expedition, Vol. 1*.
Scheeres, David, 'Thomas Feather: The One who Came Back', *Captain Scott Society Newsletter*, April 2005.
Ponting, Herbert, *The Great White South*.
Hattersley-Smith, Geoffrey & Ellen Johanne McGhie (née Gran), *The Norwegian with Scott: Tryggve Gran's Antarctic Diary 1910–1913*.
Seaver, George, *The Life of Edward Wilson, 'Birdie' Bowers of the Antarctic*.
Cherry-Garrard, Apsley, *The Worst Journey in the World*.
Scott, Captain R.F., *The Journals of Captain Scott*.

## Chapter 7 – First Role Complete

Huxley, Leonard, *Scott's Last Expedition, Vol. 1*.
Cherry-Garrard, Apsley, *The Worst Journey in the World*.
Evans, Admiral E.R.G.R., *South with Scott*.
Ponting, Herbert, *The Great White South*.

Huxley, Leonard, *Scott's Last Expedition Vol. 2: Narrative of the Northern Party.*
Davies, Frances E.C., *With Scott Before the Mast* (E.C. Davies, SPRI/MS 1267/MJ).

## Chapter 8 – New Zealand Refit and Hydrographic Surveys

Huxley, Leonard, *Scott's Last Expedition Vol. 2: Intermediate Voyages of the* Terra Nova.
Evans, Admiral E.R.G.R., *South with Scott.*
Linklater, Eric, *Voyage of the* Challenger.
Mawson, Douglas, *Home of the Blizzard.*
Huxley, Leonard, *Scott's Last Expedition Vol. 2: Reports on Biological work on* Terra Nova.

## Chapter 9 – Return to Antarctica: First Relief Voyage

Huxley, Leonard, *Scott's Last Expedition Vol. 2: Intermediate Voyages of the* Terra Nova.
Huxley, Leonard, *Scott's Last Expedition Vol. 2: Narrative of the Northern Party.*
Huxley, Leonard, *Scott's Last Expedition Vol. 2: Narrative of the Western Parties.*
Ponting, Herbert, *The Great White South.*
Evans, Admiral E.R.G.R., *South with Scott.*
Diary of Lieutenant H.L.L. Pennell, Canterbury Museum.
Pound, Reginald, *Scott of the Antarctic.*

## Chapter 10 – To Antarctica: The Final Relief

Huxley, Leonard, *Scott's Last Expedition Vol. 2: Intermediate Voyages of the* Terra Nova.
Evans, Admiral E.R.G.R., *South with Scott.*
Cherry-Garrard, Apsley, *The Worst Journey in the World.*
McElrea, Richard, & David Harrowfield, *The Polar Castaways.*
Shackleton, Ernest, *South.*

Hattersley-Smith, Geoffrey & Ellen Johanne McGhie (née Gran), *The Norwegian with Scott: Tryggve Gran's Antarctic Diary 1910–1913*.
Letter written by Thomas Williamson dated 1 February 1913, archives of the Captain Scott Society.
Campbell, David G., *The Crystal Desert*.

## Chapter 11 – The Voyage Home

Huxley, Leonard, *Scott's Last Expedition Vol. 2: Intermediate Voyages of the* Terra Nova.
Johnson, Dr Anthony M., *Scott of the Antarctic and Cardiff*.
Archives of the 4th Cardiff (St Andrews) Scout Group.
Mear, Roger, & Robert Swan, *In the Footsteps of Scott*.
Cherry-Garrard, Apsley, *The Worst Journey in the World*.
Evans, Admiral E.R.G.R., *South with Scott*.
Lashly, William, *Under Scott's Command: Lashly's Antarctic Diaries*, Ellis, A.R. (ed.), (Victor Gollancz, 1969).
Scott, Captain R.F., *The Journals of Captain Scott*.
Huxley, Leonard, *Scott's Last Expedition, Vol. 1*.
Huxley, Leonard, *Scott's Last Expedition Vol. 2: Intermediate Voyages of the* Terra Nova.
Huxley, Leonard, *Scott's Last Expedition Vol. 2: Reports on Biological work on* Terra Nova.
Poulson, Neville, & John Myres, *British Polar Exploration and Research 1818–1999: A Historical and Medallic Record*.
Huxley, Leonard, *Scott's Last Expedition Vol. 2: Outfit and Preparation*.
*Western Mail*, letter from Captain E.R.G.R. Evans RN to the editor, dated 2 May 1919.

## Chapter 12 – Newfoundland and Sealing

Report in *South Wales Daily News*, 16 August 1913.
Company records of Bowring Brothers Ltd, University of Glasgow Archives.
Shipping Reports, *Lloyds Register*, Archives of Guildhall Library, London.
Information from descendants of Commander H.L.L. Pennell RN.

Hammerton, Sir J.A., *Popular History of the Great War*, Vol. 3 (1916).
Evans, Admiral E.R.G.R., *Adventurous Life*.
Colledge, J.J., *Ships of the Royal Navy*.
*Janes Fighting Ships of World War I and II*.
Cocker, M.P., *Frigates, Sloops and Patrol Vessels of the Royal Navy*.
*Cardiff and South Wales Journal of Commerce*, Cardiff Central Library.
Keir, David, *The Bowring Story*.
Wardle, Eric C., *Benjamin Bowring and his Descendants*.
Brown, Cassie, *A Winter's Tale: The Wreck of the* Florizel.
*The Courier*, Dundee, 23 April 1919.
Ryan, Shannon, & Martha Drake, *Seals and Sealers: A Pictorial History of the Newfoundland Seal Fishery*.
Ryan, Shannon, & Larry Small, *Haulin' Rope & Gaff: Songs and Poetry of the Newfoundland Seal Fishery*.
Kean, Abram, *Old and Young Ahead: The Autobiography of Captain Abram Kean*.
England, George Allan, *Vikings of the Ice: The Greatest Hunt in the World*.

## Chapter 13 – Chartered for War Duties

Lloyd's Weekly Casualty Reports, Aug\Sept 1943, Guildhall Library, City of London.
Keir, David, *The Bowring Story*.
*Terra Nova* file, archives of Scott Polar Research Institute, University of Cambridge.
*Evening Telegraph*, 30 September 1943, St John's, Newfoundland.
England, George Allan, *Vikings of the Ice: The Greatest Hunt in the World*.
Savours, Ann, *Voyages of the* Discovery.
Archives of the Captain Scott Society.
Ryan, Shannon, & Martha Drake, *Seals and Sealers: A Pictorial History of the Newfoundland Seal Fishery*.
Squires, Harold, *SS* Eagle: *The Secret Mission, 1944–1945*.

# Bibliography

Andrieux, J.P., *East Coast Panorama: The History of Shipping Companies on Canada's East Coast* (W.F. Rannie, 1984).
Ballinger, John, *A Guide to Cardiff, City and Port* (Rees Electric Printing Works, 1908).
Brown, Cassie, *Death on the Ice: The Great Newfoundland Sealing Disaster of 1914* (Doubleday, Canada, 1974).
Brown, Cassie, *A Winter's Tale: The Wreck of the* Florizel (Flanker, 1976).
Bull, Colin, & Pat F. Wright (eds), *'Silas': The Antarctic Diaries and Memoir of Charles S. Wright* (quoting the diary of G.C. Simpson) (Ohio State University Press, 1993).
Campbell, David G., *The Crystal Desert* (Secker and Warberg, 1992).
Carvel, J.L., *Stephen of Linthouse 1750–1950* (Glasgow, 1951).
Chappell, Edgar L., *History of the Port of Cardiff* (Merton Priory Press, 1939).
Cherry-Garrard, Apsley, *The Worst Journey in the World* (Penguin Books, 1937).
Church, Ian, *Last Port to Antarctica: Dunedin and Port Chalmers: 100 years of Polar Service* (Otago Heritage Books, New Zealand, 1997).
Cocker, M.P., *Frigates, Sloops and Patrol Vessels of the Royal Navy* (Westmoreland, 1985).
Colledge, J.J., *Ships of the Royal Navy* (David & Charles, 1970).
Davies, Frances E.C., *With Scott Before the Mast* (E.C. Davies, SPRI/MS 1267/MJ).
Debenham, Frank, *Antarctica: The Story of a Continent* (Herbert Jenkins, 1959).
*Dukes County Intelligencer*, Vol. 40, No. 14. Martha's Vineyard Historical Society, MA, USA, 1998/1999).
England, George Allan, *Vikings of the Ice: The Greatest Hunt in the World* (Tundra Inc., Canada, 1924 & 1969).
Evans, Admiral E.R.G.R., *South with Scott* (Collins Press, 1921).
Fiala, Anthony, *Fighting the Polar Ice* (Hodder & Stoughton, 1907).
Fiennes, Ranulph, *Captain Scott* (Hodder & Stoughton, 2003).
Fleming, Fergus, *90 Degrees North: The Quest for the North Pole* (Granta Books, 2001).
Fuchs, Sir Vivian, *A Time to Speak* (Anthony Nelson Ltd, 1990).

Galgay, Frank, & Michael McCarthy, *Shipwrecks of Newfoundland and Labrador* (Harry Cuff, Newfoundland, 1987).

Goodenough, Clive, *History of Port Lyttelton* (1968).

Gregor, G.C., *Swansea's Antarctic Explorer: Edgar Evans 1876–1912* (Swansea City Library, 1995).

Gwynn, Stephen, *Captain Scott* (John Lane, 1929).

Hammerton, Sir J.A., *Popular History of the Great War*, Vols. 1–6 (Amalgamated Press, London, 1922).

Hattersley-Smith, Geoffrey & Ellen Johanne McGhie (née Gran), *The Norwegian with Scott: Tryggve Gran's Antarctic Diary 1910–1913* (HMSO, 1984).

Holland, Clive, *Arctic Chronology* (Garland Publishing Co., 1994).

Honeywill, Eleanor, *The Challenge of Antarctica* (Anthony Nelson, 1984).

Huntford, Roland, *Scott and Amundsen: The Last Place on Earth* (Abacus, 2000).

Huntford, Roland, *The Race for the South Pole: The Expedition Diaries of Scott and Amundsen* (Continuum, 2011).

Huxley, Leonard, *Scott's Last Expedition, Vols 1 & 2* (Smith Elder & Co., 1913).

*Janes Fighting Ships of World War I and II* (Janes Publications).

Jenkins, J. Geraint, & David Jenkins, *Cardiff Shipowners* (National Museum of Wales, 1986).

Johnson, Dr Anthony M., *Scott of the Antarctic and Cardiff* (Captain Scott Society, 1984).

Kean, Abram, *Old and Young Ahead: The Autobiography of Captain Abram Kean* (Heath Cranton Ltd, 1935).

Keir, David, *The Bowring Story*, Bodley Head, 1962).

King, H.G.R. (ed.), *Edward Wilson: Diary of the* Terra Nova *Expedition 1910–1912* (Blandford Press & Scott Polar Research Institute, 1972)

Lashly, William, *Under Scott's Command: Lashly's Antarctic Diaries*, Ellis, A.R. (ed.), (Victor Gollancz, 1969).

Limb, Sue, & Patrick Cordingly, *Captain Oates: Soldier & Explorer* (Leo Cooper, 1995).

Linklater, Eric, *Voyage of the* Challenger (John Murray, 1972).

Ludlam, Harry, *Captain Scott: The Full Story* (W. Foulsham, 1965).

McElrea, Richard, & David Harrowfield, *The Polar Castaways: The Ross Sea Party (1914-17) of Sir Ernest Shackleton* (Canterbury University Press, New Zealand, 204).

Mawson, Sir Douglas, *Home of the Blizzard* (St Martins Griffin, New York, 1998).

Mear, Roger, & Robert Swan, *In the Footsteps of Scott* (Grafton Books, 1989).

Middlemiss, Norman L., *British Shipbuilding Yards*, Vol. 3 (Shield Productions, 1995).

Moorehead, Alan, *Darwin and the* Beagle (Book Club Associates, 1978).

Ponting, Herbert V., *The Great White South* (Duckworth, 1921).

Poulson, Neville, & John Myres, *British Polar Exploration and Research 1818–1999: A Historical and Medallic Record* (Savannah Publications, 2000).

Pound, Reginald, *Scott of the Antarctic* (World Books, 1968).

Power, Rosalind, *Fort Amherst: St John's Nearest Outpost* (Jesperson Publishing, 1995).

Power, Rosalind, *A Narrow Passage: Shipwrecks and Tragedies in the St John's Narrows* (1998).

Reader's Digest, *Antarctica: Great Stories from the Frozen Continent* (Reader's Digest, 1985).

Rice, Tony, *British Oceanographic Vessels 1800–1950* (The Ray Society, 1999).

Ryan, Shannon, & Larry Small, *Haulin' Rope & Gaff: Songs and Poetry of the Newfoundland Seal Fishery* (Breakwater Books, Newfoundland, 1981).

Ryan, Shannon, & Martha Drake, *Seals and Sealers: A Pictorial History of the Newfoundland Seal Fishery* (Breakwater Books, 1987).

Savours, Ann, *Voyages of the* Discovery (Virgin Books, 1992).

Seaver, George, *The Life of Edward Wilson Birdie Bowers of the Antarctic* (John Murray, 1938).

Seaver, George, *Edward Wilson of the Antarctic* (John Murray, 1941).

Scott, Captain Robert Falcon, *Voyage of the* Discovery (Thomas Nelson & Sons, 1905).

Scott, Captain Robert Falcon, *Scott's Last Expedition: Extracts from the Personal Journals* (John Murray, 1933).

Shackleton, Sir Ernest, *South* (1921).

Sobel, Dava, *Longitude* (1995).

Squires, Harold, *SS* Eagle: *The Secret Mission, 1944–1945* (Jesperson Press, Newfoundland, 1992).

Wardle, Arthur C., *Benjamin Bowring and his Descendants* (Hodder & Stoughton, 1938).

Watson, Norman, *The Dundee Whalers* (Tuckwell Press, 2003).

Wilson, D.M. & D.B. Elder, *Cheltenham in Antarctica: The Life of Edward Wilson* (Reardon, 2000).

Wilson, Edward, *Diary of the* Terra Nova *Expedition to the Antarctic 1910–1912* (Littlehampton Book Services Ltd, 1972).

Yelverton, David, *Antarctica Unveiled* (Colorado University Press, 2001).

# Index

Abbott, PO George 155
Aberdeen 26, 31
Adams, Captain William 27, 37
Admiralty 38, 51, 54–5, 92
Admiralty Bay 177
albatross *see* Wildlife
Amorak USCGC 272
Amundsen, Roald 54, 72, 116, 157, 200

Andrassen, Mr F.L. 73
Antarctic Circle Peninsula 54
Antipodes Islands 179
Archer, Captain W. 35
Archer, W.W. 130
Arctic 259
   Smith Sound 65
   Ellesmere Island 65
   Svalbard 65
   Franz Joseph Land 69
   expeditions:
      Baldwin–Ziegler 65
      Fiala–Ziegler 66
      British Expedition 1875–76 52, 54
Akaroa Harbour 176
Angmassalik 253, 259
Atak USCGC 252, 262, 272
Atkinson, Dr E.L. 118, 129, 173–4, 187, 200
Australia 116
Azores 192
Azoic Zone 30, 167

Badcock, Captain Richard 223, 250
Baffin, William 72
Baine Johnson & Co. 44
Baldwin, Captain Evelyn Briggs 65–6

BANZARE Expeditions 266
Barbour, Captain Richard 223–4, 250
Bartlett, Captain Bob 223
Bartlett, Captain W.J. 214
Bath Iron Works 252
Batstone, Bill 101
Bay of Whales 157, 161
Beaumont, Admiral Sir Lewis 90
Beardmore Glacier 200
Beaudoin, Jonathan marine technician 275–6
bell *see Terra Nova*
birds *see* wildlife
Bishop, Captain Edwin 81, 244
Blue Glacier 173
Bluie West One 259
Bluie West Eight 259
Bowers, Lieutenant H.R. 91, 132, 138–9
Borchgrevink, Carsten 52, 54
Bowring & Son 47
Bowring Brothers Ltd 36, 44, 48, 51, 56, 72, 83, 86, 223
Bowring, C.T. & Co. Ltd 44, 50, 88, 197, 214
  insignia 49
  cod fishery, Newfoundland 35
  offices: 232
      Cardiff 50
      London 50
      Liverpool 50, 251
      New York 50
      St John's 50
      World War ship losses 49
Bowring family:
  Benjamin 41, 45, 47–8, 197
  Edward 47
  John 47, 49, 197

## Index

Charlotte 47
William 47
Charles Tricks 47–9
Eric 232
Frederick Charles, Sir 197
Bowring passenger trade 49
Bridgeman, Lady 92
Brissenden, L/Stoker Robert 177
Browning, PO Frank 155
Bruce, David & Sons (Brokers) 38, 86, 100
Bruce, Lieutenant W.M. 119, 166
Buchner, Captain Bernd 271
Buckmaster, George 193
Butter Point 155
Bull, Henrik Johan 54
Butts, Levi 241

Cabot, John 44
Caird, Sir James 27
Caledon Shipbuilding & Engineering Co. 37
Campbell, Lieutenant V.A. 91, 105, 155, 174
Camp Ziegler 70–1, 73
Canada, Confederation of 44
Cape Abruzz 70
Cape Adare 70, 157, 161
Cape Crozier 149
Cape Dillon 69, 76, 79
Cape Evans 54, 150
Cape Flora 69, 70, 76–7, 79
Cape North 157
Cape Race 218
Cape Royds 56, 84, 150, 180
Cape Town, Simons Bay 113
Cape Verde Island 105
Captain Scott Society 97, 262, 266
Cardiff, city of 92–4, 97, 195
Cardiff & South Wales Journal Commerce 217
Chafe, Levi G. 244
Champ, William S. 67, 81
Cheetham, Alfred 99, 100, 124
Cherry-Garrard, Apsley 104
Clark, Captain George 227
Clissold, Thomas 130
Coal & other services 96, 97

Coberg Island 68
Coffin, Captain Edwin B. 67–8, 70, 77, 80
Colbeck, Captain William 55, 57
Cook, Captain James 53
Cook, Frederick 52
Cope, John L. 218
Cosgrove, Thomas 56, 58
Courier 59, 219
Cox, George Addison 38
Cox, William 38
Crean, PO Tom 115, 187, 193, 198, 200
Cumberland Gulf 35
Cumming, James 35
Cutler, Ebbitt 230, 261

Dampier, William 53
David, Prof. Edgeworth 116
Davidson, Captain James 29
Davidson, Captain Robert 29
Davies, Francis E.C. 130, 159, 160–1
Davies, William E. 94
Davis, John 72
Davis, Captain J.K. 182
Dawe, Captain Charles 35, 51
Day, B.C. 157
Debenham, F. 120
De Gerlache de Gomery 54
Dennistoun, James 169
dialect, Newfoundland 234
*Discovery*, RRS 37
*Discovery* II, RRS 130
Dickason, Harry 155
Dickson, Engineer, NZ 166
*Discovery see* ships
docks:
  London 63, 87
  East India 63
  West India 63
Drake, Sir Francis 53
Drake, Francis R.H. 105
Drygalski Ice Tongue 173
Dumont de'Urville 148
Dundee Advertiser 34
Dundee Antarctic Whaling Expedition 26, 29
  Seal & Whale Fishing Co. 29
  whaling products 30

Dundee Docks 27
  jute/hemp products 30
  shipbuilding 31, 37
Dundee Heritage Trust 205
Dundee Shipbuilders Ltd 37
Dunedin 189
Duke of Abruzzi 67

Emperor penguin eggs 200
England, George Alan 228, 230, 263
Evans Cove 161
Evans, Lieutenant E.R.G.R. 88, 94, 102, 129, 173, 199–200, 209, 211
  First World War 199, 217
Evans, PO Edgar 102
Evening Telegram, St John's 257, 255
Exeter Flying Post 45

Fairweather, Captain James 29, 34
Falkland Islands (dependencies) 29, 63, 278
FALKOR rv 273–4, 276
Feather, PO Thomas 99
Fenwick, George 121
Fiala, Anthony 67–8, 72, 76, 80
Fisher, Hon. Andrew 116
Forbes, Prof. E. 168
Forde, PO Robert 198
Fort Amherst 279
Franklin, Admiral Sir John 51, 55
French Pass 177
Frobrisher, Sir Martin 72
Fuchs, Sir Vivian 266

gas mantle 28
Gibson, Joseph 38
Gilbert, Sir Humphrey 44
Glacier Tongue 150
Glengall Ironworks Co. 88, 90
Glendenning, Robert 121
Goff, W. Desmond 261, 263, 258
Gondwanaland 187
Goodyear, Captain Thomas 258
Gourlay Bros 33–4
Gran, Tryggve 105, 123–4, 129, 185
Grand Banks 45
Granite Harbour 156, 172

Great Ice Barrier 156
Greatest Hunt in the World 228, 246
Greely Relief Expedition 20
Grieve, Walter & Co. 44
Gulf of St Lawrence 215

Harbour Grace, Newfoundland 31
Harman, Captain James 41
harp/hood seals 35
Harrison, John, clockmaker 53
Harrington, Michael 257
Heald, William L. 144
Henderson, Alexander 38
Hodern, Samuel 116
Hooper, F.J. 130
Horn Head Point 215
Hudson, Henry 72
Hull 26

Jackman, Captain Arthur 51
Johansson, Captain Carl 66
Julianehaab 253, 259

Kean, Captain Abram 51, 81, 218, 220, 223, 225, 228, 230
Kean, Captain Westbury 223, 250
Kean, Captain Jacob 250
Keohane, PO P. 174, 198,
Kemp, Ebenezer 34
Kennedy, Captain N.J. 214, 217–18
King Edward VII Land 156
King George V 203
Kingsbridge Devon 45
Kinsey, Sir James 120, 176
Kjeldsen, Captain Johan Kjeld 66, 71–2, 75, 78
Knee, Captain Job 244

Labrador, seal/cod fishing 42, 44–5
Lashly, PO William 129, 198–200
Laurel USCGC 165, 171, 255, 273
Levick, Murray 155
Lillie, D.G. 166, 170, 203–4
Lloyds Register:
  Weekly Index 216
  reports 252, 263
Lucas sounding machine 167
Lucas, Francis 169

## Index

Lush, Captain Llewellyn 257, 263
Lyttelton, New Zealand 62, 120–1

Macquarie Island 165
Madeira, Funchal 103, 105
Marie, Jean (marine technician) 275–6
Markham, Admiral Sir Albert 90
Markham, Sir Clements 52, 95
Marsh McLennan 50
Mather, PO John 173
Mawson, Sir D. 16
Mead, Captain G.H. 266
Meares, Cecil H. 119, 172
Melbourne 116
Monitou USCGC 273
Mount Melbourne 173
Mill, John & Co. 121
Miller, H.J. 118
McCarthy, Mortimer 147
MacGavin, Robert 38
MacKay, Captain Harry 38, 56–8
MacKenzie, Sir Alexander 72
MacKenzie, Captain K.N. 266
McMurdo Sound 60, 83–4, 149–50
Moore, Mr/Mrs William 121
Motor sledges 200
'Moses' (dog) 68
mules 169
Munth/Mount, Dr 76
Munn, John & family 218
Myre, Segurd 77

Nares, Captain G.S. 168
Narsarssuak 259
Neale, W.H. 130
Nelson, Edward W. 209
Newcastle 26
Newfoundland Base Contractors 250
Newfoundland return 213
Newnes, Sir George 54
New Trinidad Island 109
New Zealand expedition
  surveys/reliefs 164
'N★★★★r' (cat) 117, 171–2, 175
Northwest Passage 52, 72
Norwegian Church, Cardiff 269

Oates, Captain L.E.G. 101, 129
Oamaru 187
Observation Hill 159, 182, 203
Olive Block fossils 187
One Ton Depot 200, 202

Panmure shipyard 32–3, 56
Parry, Sir William 72
Parsons, Captain John 213, 250
Paton, Captain James 213
Paton, James 213
Peary, Robert 52
Pennell, Lieutenant H.L.L. 91, 110, 112,
  150, 177, 187, 199, 216
Penny, Edward J. 232
Peterhead 26
Pianoforte 90
Plimsoll line 92
Ponting, Herbert 120–1, 126, 175,
Port Chalmers 121, 123, 189
Priestley, R.E. 120, 155

Radcliffe, Daniel 95, 97, 101, 198, 211
Red Cross Line 49, 218
Rennick, Lieutenant E. de P. 91, 149, 157,
  199
Richards, Sir George 169
Ridley Beach 159
Ridley & Co. 44
Rio de Janiero 192
Riley, Lieutenant E. 116
RNVR volunteers 88
Robertson Bay 157, 159, 172
Robinson, Captain Thomas 29
Rolley, Leighton (marine technician) 271,
  275
Ross, Captain James Clark
Ross Harbour, Auckland Islands 60, 62
Ross, Sir John 72
Royal Geographical Society 52
Royal Hotel, Cardiff 97, 195
Royal Society 52, 55
Royal Yacht Squadron 92, 102
Rudolf Island 67
RV Falkor (Schmidt Ocean Institute) 271

*Saturday Evening Post* 228
Schmidt Ocean Institute 271

Schmidt scientific parties 273
Scilly Isles 192, 194
Scott, Kathleen (née Bruce) 83
Scott, Robert Falcon:
   first expedition (Discovery) 52, 55
   Farthest South 83
   Voyages of the *Discovery* 83
   second expedition (Terra Nova) 81, 83, 199, 200
   British Expedition Committee 1910 81, 85–6
   relationship with Shackleton 85
Scout Association 100–1, 266
Scout Groups:
   4th Cardiff (St Andrews) 100–1,192
   Monowai Sea Group (Terra Nova) NZ 187
   Oamaru Group (Scott's Own) NZ 189
Seals *see also* wildlife 224
   Newfoundland hunt 219
   Phoca groenlandica 235
   Cystophora cristata 235
Shackleton, Sir Ernest 83, 94
   Imperial Trans-Antarctic Expedition 85
Ships (other):
   *Atak* 233, 255, 271
   *Active* 29
   *Adventure*, HMS 53
   *Alert*, HMS 168
   *America* 65–6, 68, 71
   *Amethyst*, HMS 199
   *Antarctic* 54
   *Arctic* 35
   *Aurora* 33, 35, 51, 85, 181, 216, 226, 276
   *Balaena* 29
   *Bantam Cock* 103
   *Beagle* 167
   *Bear (of Oakland)* 33
   *Belgica* 66
   *Beothic* 226
   *Blanche*, HMS 160
   *Bloodhound* (HMS *Discovery*) 168
   *Broke*, HMS 199, 217
   *Challenger*, HMS 30, 167
   *Charlotte* 46
   *Colonel F. Armstrong* 252
   *Commodore* 44

*Corinthic*, RMS 113, 116
*Devonia* 102
*Diana* 30, 33, 215, 241, 244,
*Discovery*, RRS 30–1, 52, 55–6, 60
*Discovery II* 161
*Eagle* 46, 224, 245, 277
*Eagle II* 230, 263
*Endeavour*, HMS 53
*Endurance* 85, 277
*Erebus*, HMS 28, 54, 148
*Erik* 215, 277
*Esquimaux* see *America* 35, 65, 277
*Exmouth*, HMS 160
*Falcon* 95
*Florizel* 218, 226
*Fram* 157, 158
*Fitzroy* 15
*Frithjof* 65, 71–2
*Gjoa* 72
*Gordon C.* 218
*Greenland* 227
*Hawk* 44, 218
*Hogue*, HMS 199
*Home* 218
*Inverclyde* 105
*Invincible*, HMS 92
*Irresistible*, HMS 199
*Jean of Bo'ness* 38, 41
*John Biscoe* 15
*Kite* 277
*Laurel* USCGC 254
*Lobelia*, HMS 214
*Lyttelton* 120
*Maori* 118
*Margaret Jane* 47
*Mastiff* 44
*Maud* 37
*Mercedes* 93, 192
*Morning* 55–7, 60–1, 83, 277
*Mutine*, HMS 113
*Nascopie* 227
*Nelson* 192
*Newfoundland* 227, 239
*Neptune* 33, 240
*Nimrod* 33, 83, 99, 277
*Pandora*, HMS 113
*Plover*, HMS 44
*Plucky* 123

*Polar Star* 29
*Powerful*, HMS 116
*Prospero* 218
*Pretext*, HMS 15
*Queen Mary*, HMS 199, 216
*Ranger* 33, 224, 277
*Ravenswood* 102
*Resolution*, HMS 53
*Sagona* 240, 244
*Sandhurst*, HMS 160
*Saxon*, RMS 113
*Scotia* 15, 277
*Southern Cross* 54, 227, 240, 277
*Southern Quest* 195
*Stephano* 226–7
*Thetis* 33–4, 71, 226, 241, 277
*Terror*, HMS 28, 54, 148
*Triton* 179
*Velocity* 47
*Victory III*, HMS 161
*Viking* 215, 223, 277
*Warrimoo* 118
*William Scoresby* 161
*Worcester*, MTS 99
*Wolf* 44, 226
salted cod industry 45
Schmidt Ocean Research 271, 273–4, 276
SHRIMP equipment 275
Simons & Co., William (Clyde) 217
Simpson, George C. 106, 173
Skelton AB, James 199
Smith, Reginald 90
South Pole 250
*South Wales Daily News* 213
Spencer, Prof. 116
Speyer, Sir E.D. 85
Stephen, Andrew Henderson 38
St John's, Newfoundland 82, 224
  Port Authority 257
Stephen, Alexander & Sons Ltd 30–1
  builders, *Terra Nova* 30–1
  Alexander 33, 37
  William 33, 37
  Alice 35
  as whaling company 31
  processing plant/activities 35
  Shipyard Panmure, Dundee 31
  River Tay 31

yards on Clyde 33–4
yards at Burghead, Moray Firth 30–1
Stewart Island NZ 63
St Paul Island 115
Sturrock, George William Lyon 38
Swenson, Lieutenant F. 266
Swensen, Gus 266

Tabarin, Operation 278, 280
Tasman, Abel 53
Tasmania 56
Taylor, Griffith 120, 129
Terra Nova Bay 173
Terra Nova House 49
Terra Australis Incognita 53
*Terra Nova*, SS
  design 33
  launch 34
  ownership changes 38
  registration documents 39
  sold to Ziegler 72
  Admiralty 55
  British Expedition 52, 55
  expedition hull plans 66, 86
  engine power 33–4
  compass/binnacle 88, 270
  fitting out 86
  fire aboard 106
  pumps 136–8, 140–1, 165
  resold to Bowrings 43
  ship's bell 73–4, 199, 243
  storm 128
  rigging 38
  home port, Dundee 33
  as relief ship 74
  Antarctic expedition ship 83
  Arctic 75
  return to Newfoundland 81, 213
  refit 87, 217, 249
  song 157, 220
  sealing records/rates 238
  First World War Atlantic crossing 217
  foundering/loss 262
Taylor-Griffith 181
'Te Han' 120
Teplitz Bay 67–8
Thompson, Captain F.T. 168
Three Kings Island 169

Thompson, Alexander Gordon 37
Tromsø, Norway 73, 80
Tundra Books Ltd 229

US Coastguard 262

Van Diemen's Land/Tasmania 53
Vardo, Norway 67, 69
Vikings of the Ice 228

Webster of Faversham 41
Wellman, Walter 66
Welsbach, Carl Von 28
*Western Mail* 195, 211
whales
  *Balaena mysticetus* 28, 29
  the 'Right' whale 28
  baleen whalebone 28
  blubber 206
  *Balaenoptera sibbaldi* 207
  *Balaenoptera rostrata* 207
  *Balaenoptera borealis* 206

killer whale 207
*Megaptera longimana* 206
wildlife 108, 126, 148
Williams, Engineer William 111, 129–30, 165
Williamson, PO Thomas 154, 185
Winsor, Captain S.R. 217
Wilson, Dr E.A. 88, 97, 106
Wood Bay 170
Woods Hole Oceanographic Institution 275
wreck, finding of 271
Wright, Charles S. 116
Wyman & Sons 196

Yule, Captain Charles 27

Zenith Dredging Co. (Laurel USCGC) 254
Ziegler, William 65, 71–2
  bought *Terra Nova*